THEMES IN
geography

International Issues

Calvin Clarke

Hodder & Stoughton
LONDON SYDNEY AUCKLAND TORONTO

British Library Cataloguing in Publication Data
Clarke, Calvin
Study themes in geography: international issues.

1. Human geography
I. Title II. Series
304.2

ISBN 0 340 531630

First published 1991

Typeset by Litho Link Ltd, Welshpool, Powys, Wales.
Printed in Hong Kong for the educational publishing division of Hodder and Stoughton Ltd, Mill Road, Dunton Green, Sevenoaks, Kent by Colorcraft.

Contents

Acknowledgements

The author and publishers thank the following for permission to reproduce photographs in this book (the first number refers to the page, the number in brackets to the column):

Adrian Arbib/Impact Photos 46(2), 46(3)
Asmin Khan/Oxfam 60(1)
Aspect Picture Library 60(2&3)
Associated Press/Topham 43(3), 89(3)
B and C Alexander 53(2)
Bill Wise/Oxfam 39(1&2), 60(1)
Christian Aid Photo Library 26(2)
David Horwell Photography 42(2)
Ecoscene/Pearson 58(1)
Ecoscene/Sally Morgan 78(3)
Frank Lane Picture Agency 53(1), 78(3)
J. Allan Cash 11(1), 19(2), 21(3), 46(1)
Kim Westerskov/Oxford Scientific Films 53(1)
Marc and Evelyne Bernheim 69(2)
Mark Edwards/Still Pictures 59(1)
Medical Illustration/St. Bartholemews Hospital 62(1)
Oxfam 37(1&2), 60(3)
Oxfam/J. Hartley 58(2), 59(2)
Penny Tweedie/Impact Photos 19(3)
Piers Cavendish/Impact 38(2)
Popperfoto 41(3)
Robert Frerck/Susan Griggs Agency 49(1&2)
Robert Harding Picture Library 11(3), 22(1), 31(2), 41(1&2), 48(1&2), 78(3)
Robert Nicholls/Oxfam 62(2)
Seaphot Ltd: Planet Earth 67(2)
Susan Griggs Agency 58(3)
World Health Organisation 62(1)

Introduction

This book has been designed for pupils of all abilities to learn key ideas 6 and 12-17 of the Scottish Standard Grade syllabus. Each unit of work is fully differentiated and is divided into several sections:

1 All pupils read the CORE TEXT

2 All pupils answer the CORE QUESTIONS

Pupils then choose either

3 FOUNDATION QUESTIONS

or

4 GENERAL QUESTIONS

or

5 EXTENSION TEXT, EXTENSION QUESTIONS AND CREDIT QUESTIONS

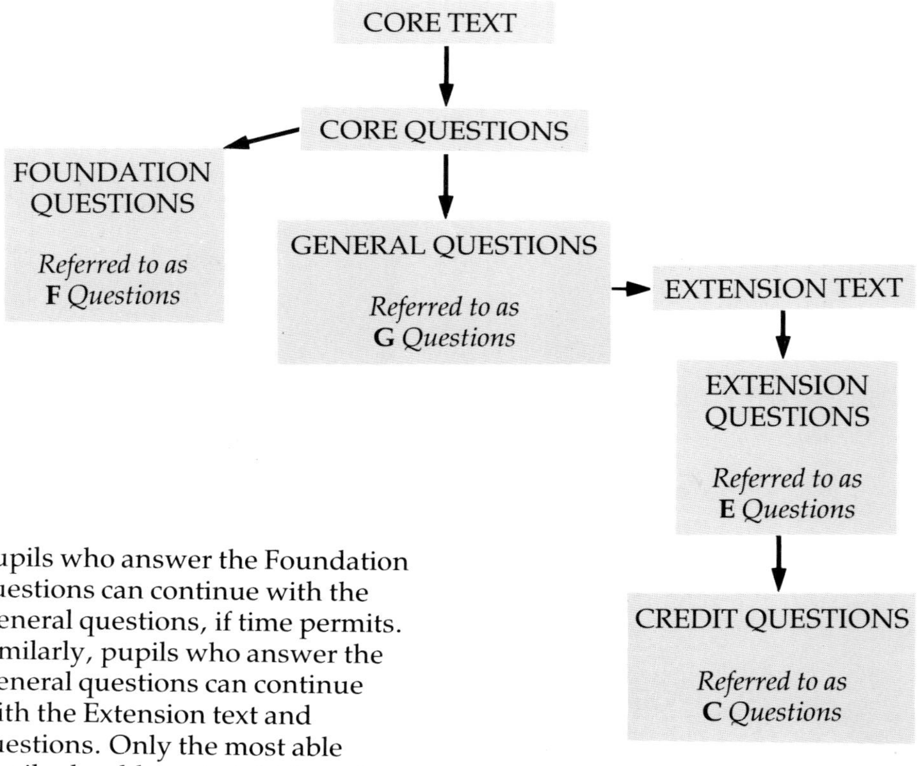

CORE TEXT
↓
CORE QUESTIONS
↙ ↓
FOUNDATION QUESTIONS *Referred to as* **F** *Questions*
GENERAL QUESTIONS *Referred to as* **G** *Questions*
→ EXTENSION TEXT
↓
EXTENSION QUESTIONS *Referred to as* **E** *Questions*
↓
CREDIT QUESTIONS *Referred to as* **C** *Questions*

Pupils who answer the Foundation questions can continue with the General questions, if time permits. Similarly, pupils who answer the General questions can continue with the Extension text and questions. Only the most able pupils should go straight from the Core questions to the Extension text.

This method of differentiation is used in all units except for those that involve investigations (units, 3, 7, 11 and 15).

For the groupwork exercises in units 9 and 13, it is necessary for the teacher to prepare the answers in advance.

Assessable Elements

The Core questions and Extension questions in each unit test mostly Knowledge and Understanding of the Key Ideas. The Foundation, General and Credit questions test mostly Evaluating skills through case studies.

Learning Approaches

In units 5, 7, 9 and 13, there is an additional section of Core Groupwork. The exercises are placed at the end of the units and have been designed so that pupils of all abilities can fully participate in them.

Main Ideas

Unit 1 Population facts are gathered in different ways with varying degrees of accuracy.

Distribution can be analysed in terms of population totals.

Unit 2 Measured social characteristics include various indicators of standards of living.

Distribution can be analysed in terms of population densities.

Unit 3 Distribution of population can be related to economic, political and environmental factors.

Higher population densities become possible with more advanced technology.

Unit 4 Distribution of population can be related to economic, political and environmental factors.

Unit 5 Measured social characteristics include birth rate and death rate.

Unit 6 Variations in birth and death rates can be related to social and economic influences.

Population change can create economic problems.

Unit 7 Population change is influenced by immigration and emigration.

Population change can create economic, social and political problems.

Unit 8 Population change is influenced by immigration and emigration.

Population change can create economic, social and political problems.

Unit 9 In the tropical rainforests, human activity can trigger short- and long-term problems.

In the developing world, a variety of problems are caused by human factors.

Unit 10 The explorations of the oceans can trigger short- and long-term problems.

Unit 11 Human activity on the land can lead to the spread of deserts.

In the developing world, a variety of problems are caused by physical and human factors.

Unit 12 In the developing world, a variety of problems are caused by physical factors.

Unit 13 Patterns of trade illustrate the interdependence of different parts of the world.

Interdependence has political and economic implications.

In the developing world, a variety of problems are caused by human factors.

Unit 14 Different forms of self-help and international aid are required to meet a variety of problems.

Unit 15 Certain countries are dominant in international relations, in particular the USA and USSR.

Unit 16 Certain countries are dominant in international relations. Their influence in conflicts is a reflection of their location, size, population, resource base and level of technology.

Unit 17 Certain alliances are dominant in international relations.

Patterns of trade illustrate the interdependence of different parts of the world.

Unit 18 The European Community is a dominant alliance in international relations.

Patterns of trade illustrate the interdependence of different parts of the world.

Assessable Elements

Unit	Key Ideas*	Knowledge + Understanding (F/G Levels)			Knowledge + Understanding (G/C Levels)			Evaluating F Level				Evaluating G Level				Evaluating C Level				Investigating	Group Discussions
		a	b	c	a	b	c	a	b	c	d	a	b	c	d	a	b	c	d		
1	13/12	✓	✓	✓		✓		✓	✓	✓		✓	✓	✓	✓	✓	✓	✓	✓		
2	13/12	✓	✓	✓		✓	✓	✓	✓	✓	✓	✓		✓	✓	✓	✓				
3	12		✓																	✓	
4	12	✓	✓	✓		✓		✓		✓	✓	✓	✓	✓		✓	✓		✓		
5	13/14	✓	✓	✓		✓	✓	✓	✓	✓	✓	✓	✓	✓		✓	✓	✓			✓
6	14/17	✓	✓	✓		✓		✓	✓	✓		✓	✓	✓		✓	✓	✓	✓		
7	14/17		✓									✓		✓			✓	✓		✓	✓
8	14/17	✓	✓		✓	✓		✓	✓	✓		✓	✓	✓		✓	✓	✓			
9	6/17		✓			✓		✓	✓	✓	✓	✓	✓	✓	✓		✓		✓		✓
10	6	✓	✓			✓		✓	✓	✓		✓	✓	✓	✓	✓	✓	✓			
11	6/17	✓				✓		✓	✓			✓				✓				✓	
12	17	✓	✓			✓			✓	✓	✓	✓	✓	✓		✓	✓	✓	✓		
13	16/17	✓	✓			✓		✓	✓	✓		✓	✓	✓		✓	✓	✓			
14	17	✓	✓			✓		✓	✓	✓	✓	✓	✓	✓		✓	✓	✓			✓
15	15	✓	✓	✓			✓	✓				✓				✓				✓	
16	15	✓	✓		✓	✓		✓	✓	✓		✓	✓	✓		✓	✓	✓			
17	15/16	✓	✓	✓		✓		✓	✓	✓		✓	✓	✓		✓	✓	✓			
18	15/16	✓	✓		✓	✓		✓	✓	✓	✓	✓	✓	✓		✓	✓	✓	✓		

*Numbers refer to revised 1991 Key Ideas.

UNIT 1 Counting the People

Core Text

1A Population

The number of people living in a country or region is called its **population.**

The population of Scotland is 5 million. Scotland is part of the United Kingdom, which has a population of 57 million. The United Kingdom is part of the continent of Europe, which has a population of 500 million. And Europe is part of the World, which has a population of 5000 million.

1B A Census

A census is a population count to find out how many people there are in a country or region. In Britain we take a census every 10 years. The last one was in 1991. Every household was given a CENSUS FORM to fill in, on exactly the same date – Sunday 21 April. The forms were then collected and the figures added up.

The census has questions on each person in every house, to find out:

(*a*) **Basic facts** – age, sex, nationality

(*b*) **Extra facts** – other useful details such as occupation, type of house, language spoken.

1C Census Problems

People have to fill in census forms or they may be fined. But many countries still find it difficult to take an accurate census. These are some of the reasons:

(*a*) they are very expensive,

(*b*) it is difficult to reach some villages,

(*c*) some people cannot read or write,

(*d*) it is difficult to count people if there is a war,

(*e*) some people are nomads – they have no fixed home,

(*f*) some people do not want to tell the truth.

1D Countries with the Largest Populations

The following are the countries with the largest populations (in millions):

USA	242M
Bangladesh	101M
USSR	280M
China	1072M
Indonesia	167M
Nigeria	98M
Japan	121M
India	766M
Pakistan	99M
Brazil	138M

1E Why Censuses Are Taken

Every government needs to know how it should spend its money and where it should spend its money. So it needs to know how many people live in different regions, as well as such things as their age, sex, job, language and type of accommodation. The information about the population of a country helps a government in planning how to spend its money. For example census data helps in working out how much money is needed for pensions, schools and hospitals and where factories and houses should be built.

1F Population Pyramids

The diagram below shows the population pyramid for the world.

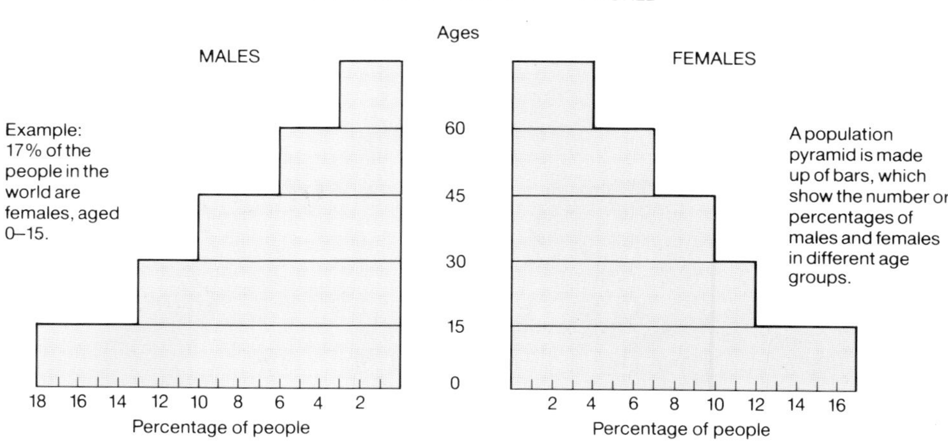

Core Questions

RANK	COUNTRY	POPULATION (Millions)
1	China	1072
2		
3		
4		
5		
6		
7		
8	Bangladesh	101
9		
10		

Look at 1D.

1 Complete a copy of the table above by ranking (putting in order) the 10 countries with the largest populations in the world.

Look at 1B.

2 What is a census?

3 Which three of the following does a census find out about people?

(*a*) their age (*b*) their sex (*c*) their weight (*d*) their hobbies (*e*) their occupation (*f*) their make of car.

Look at 1C.

4 Why do poor countries find it difficult to take censuses?

5 Why are nomads difficult to count?

6 Why should a census not be taken when there is a war?

Look at 1E.

7 Why do governments need to know how many children and how many old people there are in the country?

Look at 1F.

8 How many people in the world are

(*a*) 0 – 14 year old males

(*b*) 15 – 29 year old females

(*c*) 30 – 44 year old males

(*d*) 45 – 59 year old females

(*e*) males aged 60 or over?

Questions

Case Study of Ethiopia

Look at fig. 1.1.

F1 Which age group in Ethiopia has

(*a*) the most males and females,

(*b*) the fewest males and females?

F2 Which age groups have more females than males?

Look at fig. 1.2.

F3 Which two of the facts in fig. 1.2 best explain why censuses in Ethiopia are often wrong?

A new report claims only 4 out of every 100 Ethiopians can read and write

F4 What is the headline above trying to say about the problems of taking censuses in Ethiopia?

'Ethiopia should not have a census, it's a waste of money.'

F5 Do you agree with the statement above?
Give reasons for your answer.

Ethiopian Ministers meet to work out how many schools and teachers they will need in the next 5 years

Look at the headline above.

F6 Which is more important for the government of Ethiopia to know about its people – their ages or the jobs they do?
Give a reason for your answer.

Questions

Case Study of Malaysia

Look at fig. 1.5.

G1 Compare the population in different age groups in Malaysia.

G2 Compare the number of males and females in each age group in Malaysia.

Look at figs. 1.4 and 1.6.

G3 Do you think that censuses in Malaysia will be accurate?
Give reasons for your answer.

Malaysia Ministers clash over the next census – is it worth it?

Look at the headline above.

G4 Give one argument for and one argument against Malaysia holding another census.

G5 What is the cartoon above trying to say about the reasons for taking a census?

"Taking a census should not be expensive. We just hand out a census form to each household and collect it later." (Malaysian officer)

G6 Explain how the above statement is exaggerated.

Resources

Case Study of Ethiopia

Ethiopia: Facts

1 Population 46 million (1986)
2 20 million children
3 Many nomads
4 ¾ of the people are farmers
5 Civil war going on
6 Hot dry climate

Fig. 1.2

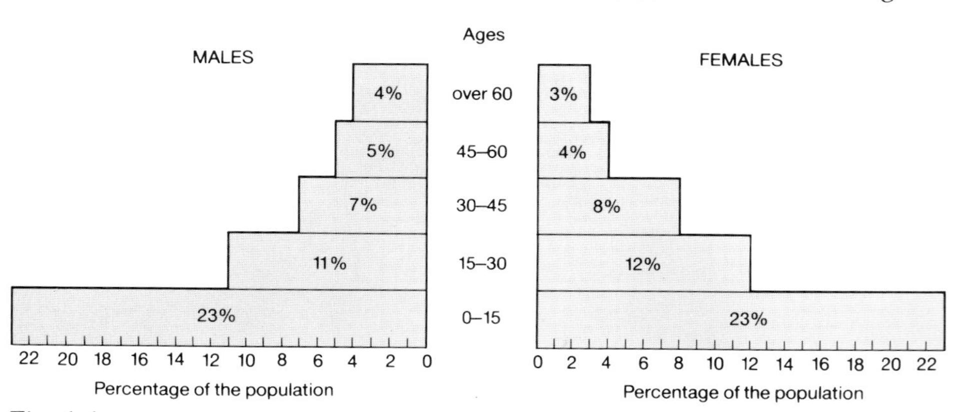

Fig. 1.1

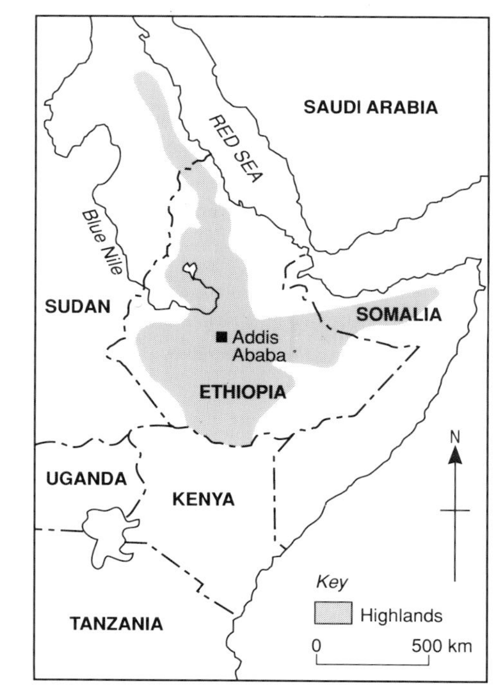

Fig. 1.3

Case Study of Malaysia

Malaysia: Facts

1 Population 17 million (1988)
2 70% is covered with dense rainforest
3 Mountains inland with few roads
4 Some nomads live in the rainforest
5 Most people live near the coast

Fig. 1.4

Malaysia: People

6 million Malays speak Malaysian
4 million Chinese speak Hokkien or Chinese
1 million Indians speak Tamil or Urdu
Many other races
40% of the people cannot read or write

Fig. 1.6

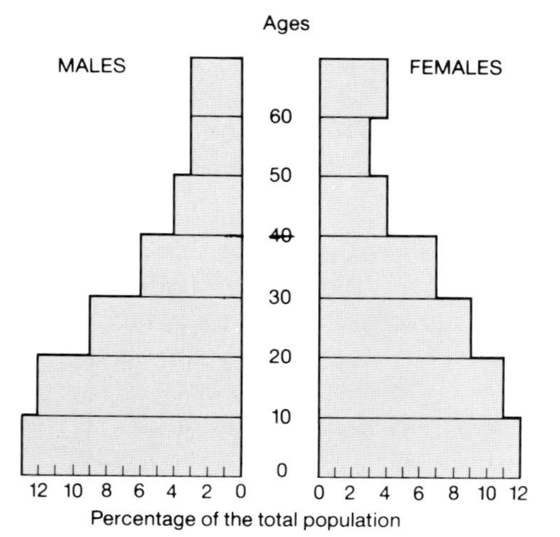

Fig. 1.5

Case Study of Nigeria

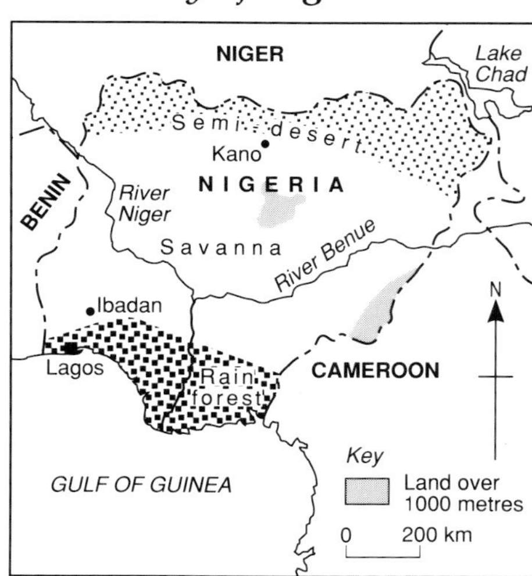

Fig. 1.7

Nigeria: Facts

1 Population 105 million (1988)
2 Most people live in the south
3 The north-east is very remote with few roads
4 Many languages spoken
5 34% of adults are literate

Fig. 1.8

Nigeria: Census Results and Estimates

1952	30 million	1973	79 million
1962	45 million	1974	71 million
1963	56 million	1975	84 million
1969	64 million	1976	77 million
1970	56 million		

Fig. 1.9

There is a movement of population in Nigeria away from the overcrowded south to the less densely populated north.

The number of people in each region in Nigeria determines the money they receive from the Government and the number of votes they have in Parliament.

Fig. 1.10

Extension Text

1G Census Problems

In countries such as Britain censuses are quite reliable. But they are only taken every 10 years and a lot of population changes can take place between censuses. It also takes several years for all the information to be analysed even with the help of computers. Some countries such as Sweden have now stopped taking censuses. They believe there are better ways of counting the people.

1H Vital Registrations

A census is not the only way of counting people. Records can also be kept of vital events such as births, deaths, marriages, adoptions, divorces.

In Britain these **vital registrations** are compulsory. Without them people would find it very difficult to get passports, health certificates, life insurance and other things. This makes them a reliable way of counting people. In poorer countries they are unreliable because they are not compulsory, many people are illiterate (cannot read or write) and live a long way from the registration offices.

Vital registrations provide a continuous record of population changes and are quite cheap. A census on the other hand, records details of the population at one point in time and is expensive to undertake.

1J Sample Surveys

Instead of getting facts from everyone in the country, a sample number of people can be asked eg 10 per cent. The results are not quite as reliable and they have the same problems as a full census. But they are cheaper and quicker to do.

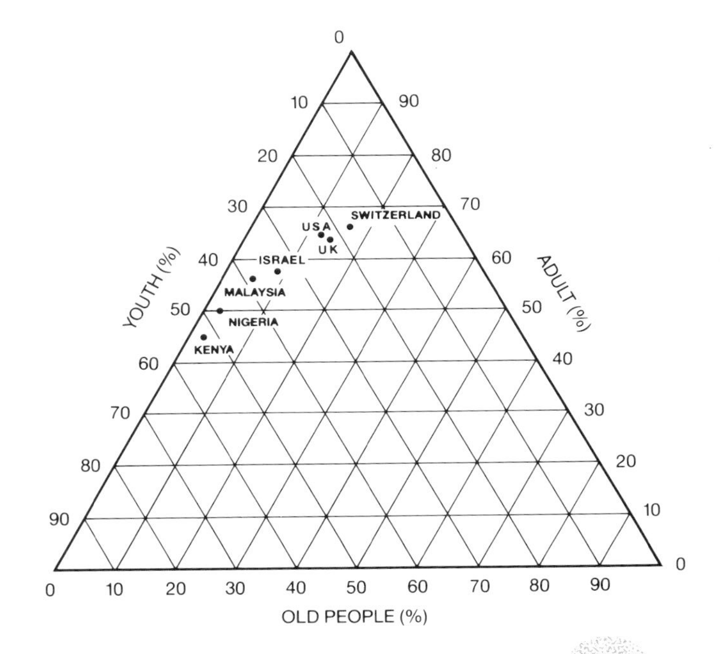

1K Triangular Graphs

The number of people for each age range and sex in a population is called the **population structure**. It can be shown by a population pyramid. Also, if the population is divided into youths, adults and old people, the population structure can be shown on a triangular graph.

Questions

Look at the Extension Text.

E1 Explain why some countries have stopped taking censuses.

E2 What are 'vital registrations'?

E3 Explain why vital registrations in poorer countries are unreliable.

'A census is like a photograph. Vital registrations are like a film'.

E4 Explain what the statement above means.

E5 What is a sample survey of the population?

E6 Explain what is meant by 'population structure'.

Questions

Look at the triangular graph in 1K.

C1 Compare the population structure of Nigeria and the UK.

Look at figs. 1.8 and 1.9.

C2 Do you think that the population totals in fig 1.9 accurately show the changes in population in Nigeria?

Give detailed reasons for your answer.

A census, a sample or vital registrations? Nigerian Ministers fail to agree on the best method of counting people

Look at the headline above.

C3 Explain the different points of view that Nigerian ministers would have.

Look at fig. 1.10

C4 The following statement is biased.

'Nigeria should not count its population for at least another 10 years. It is a waste of public money.' (South Nigerian politician).

Explain how it is biased and suggest reasons why.

UNIT 2 Standard of Living and Population Density

Core Text

2A *Standard of Living*

Censuses tell us how many people live in a country. They can also tell us how well-off the people are. This is called their **standard of living**. Most people in Ethiopia are not well-off. They have a **low standard of living**. Most people in Scotland are well-off. They have a **high standard of living.**

2B *How to Work Out Standard of Living*

It is difficult to work out how well-off the people in a country are. Some of the indicators which are often used are shown in the table.

WEALTH	THE AVERAGE INCOME PER PERSON This is the average amount of money each person receives during the year.
FOOD EATEN	THE AVERAGE NUMBER OF CALORIES PER PERSON This is the average amount of calories each person eats every day.
HEALTH	THE NUMBER OF PEOPLE PER DOCTOR
EDUCATION	THE NUMBER OF PEOPLE WHO CAN READ AND WRITE (ARE LITERATE)
INDUSTRIALIZATION	THE PERCENTAGE OF PEOPLE IN AGRICULTURE

2C *Developed and Developing Countries*

Countries with a high standard of living are called **Developed Countries**. They are sometimes called '**The North**' because they are found mainly in the northern hemisphere. Countries with a low standard of living are called **Developing Countries**. They are sometimes called '**The South**' or '**The Third World**'.

Examples of Developed Countries are USA, USSR, France and Great Britain, while Argentina, Saudi Arabia and Ethiopia are all part of the Developing World.

2D *Population Density*

A census can also tell us how crowded a country is. This is called the **population density.**

$$\textbf{Population density (people per sq km)} = \frac{\textbf{The population of a country}}{\textbf{The area of the country (in sq km)}}$$

England and Czechoslovakia have approximately the same area. But England has three times as many people. So England is more crowded. It has a **high population density**. Czechoslovakia is less crowded. It has a **lower population density.**

2E *Dot Distribution Maps*

The maps of England and Czechoslovakia are **dot distribution maps**. Each dot stands for 1 million people. They show the population density clearly. The dots are placed where most of the people live. So they also show the **population distribution** – where the people in a country actually live.

Core Questions

Look at 2A.
1 If a country has a high standard of living, what does this mean?

Look at 2B.
2 Which three of the following can be used to work out a country's standard of living:
(*a*) the population
(*b*) the wealth
(*c*) the area
(*d*) the number of calories each person eats
(*e*) the number of people per doctor
(*f*) the number of Olympic gold medals they win?

Look at 2C.
3 What is a Developed Country?
4 What is a Developing Country?
5 What is another name for a Developing Country?
6 Name three Developed and three Developing Countries.
7 Is the area shown in this photograph densely populated?

	Population (millions) mid 1980s	Area (millions of sq km)	Population Density (per sq km)
Argentina	30	3	10
Mexico	80	2	
Canada	27	9	
Thailand	50	½	100
Sudan	22	2	
Egypt	50	1	

8 Copy and complete the table above by working out the population density, using the formula:

$$\text{Population density} = \frac{\text{Population}}{\text{Area}}$$

Questions

Case Study of Scotland and Liberia

Look at fig. 2.1.
F1 Where do most people in Scotland live?
F2 Where do the fewest people in Scotland live?
F3 Where do most people in Liberia live?
F4 Where do the fewest people in Liberia live?

'Fig. 2.2 shows that all of Scotland is more densely populated than Liberia'
F5 Explain in what way the statement above is exaggerated.

Look at fig. 2.3.
F6 Which country – Scotland or Liberia
(*a*) is richer?
(*b*) has better fed people?
(*c*) has healthier people?
(*d*) has better educated people?
F7 Which country – Scotland or Liberia has the higher standard of living? Give reasons for your answer.

Sketch of part of Liberia

Look at the sketch diagram above.
F8 The average income in Liberia is only £300 per year. Do you think that everyone in Liberia is poor? Give reasons for your answer.

Look at figs. 2.3 and 2.4.
'Liberia's standard of living has improved since 1960' (Liberian minister)
F9 Do you agree with the statement above? Give reasons for your answer.

Questions

Case Study of Argentina and Saudi Arabia

Look at fig. 2.5.
G1 Describe the population distribution in Argentina.

G2 Describe the population distribution in Saudi Arabia.

Look at fig. 2.6.
'The photograph above shows that Saudi Arabia has a high population density'
G3 Explain in what way the above statement is misleading.

Look at fig. 2.7.
'Saudi Arabia has a higher standard of living than Argentina'
G4 Give one argument for and one argument against the point of view above.

Look at figs. 2.7 and 2.8.
G5 Describe how the standard of living in Argentina changed between 1960 and 1980.

'Fig. 2.7 shows that everyone in Saudi Arabia has a high income.'
G6 Explain in what way the statement above is exaggerated.

Resources

Case Study of Scotland and Liberia

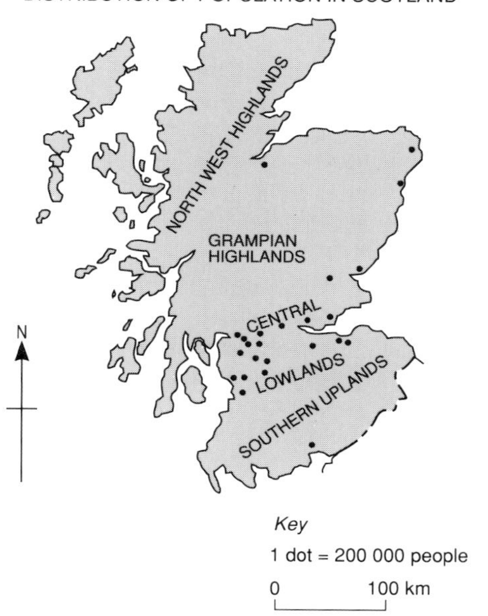

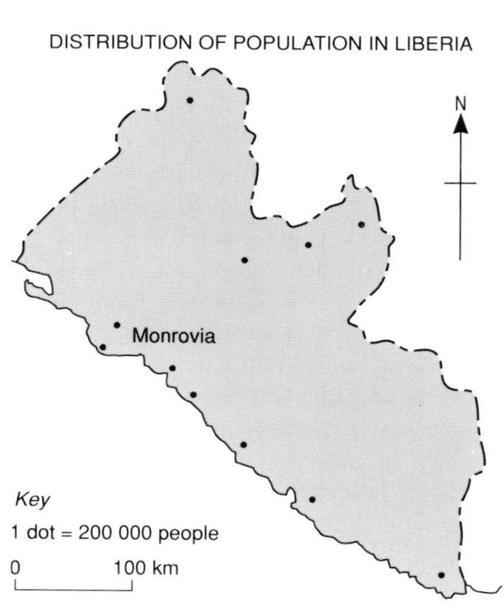

Fig. 2.1

	Population (thousands) 1980	Area (thousands of sq km)	Population Density
Scotland	5200	80	65
Liberia	2000	100	20

Fig. 2.2

	Scotland (1980)	Liberia (1980)
Average income per person	£4000	£300
Calories per day	3100	2200
Patients per doctor	650	8600
People that can read and write	99%	34%

Fig. 2.3

	Liberia (1960)
Average income per person	£200
Calories per day	2200
Patients per doctor	12000
People that can read and write	18%

Fig. 2.4

Case Study of Argentina and Saudi Arabia

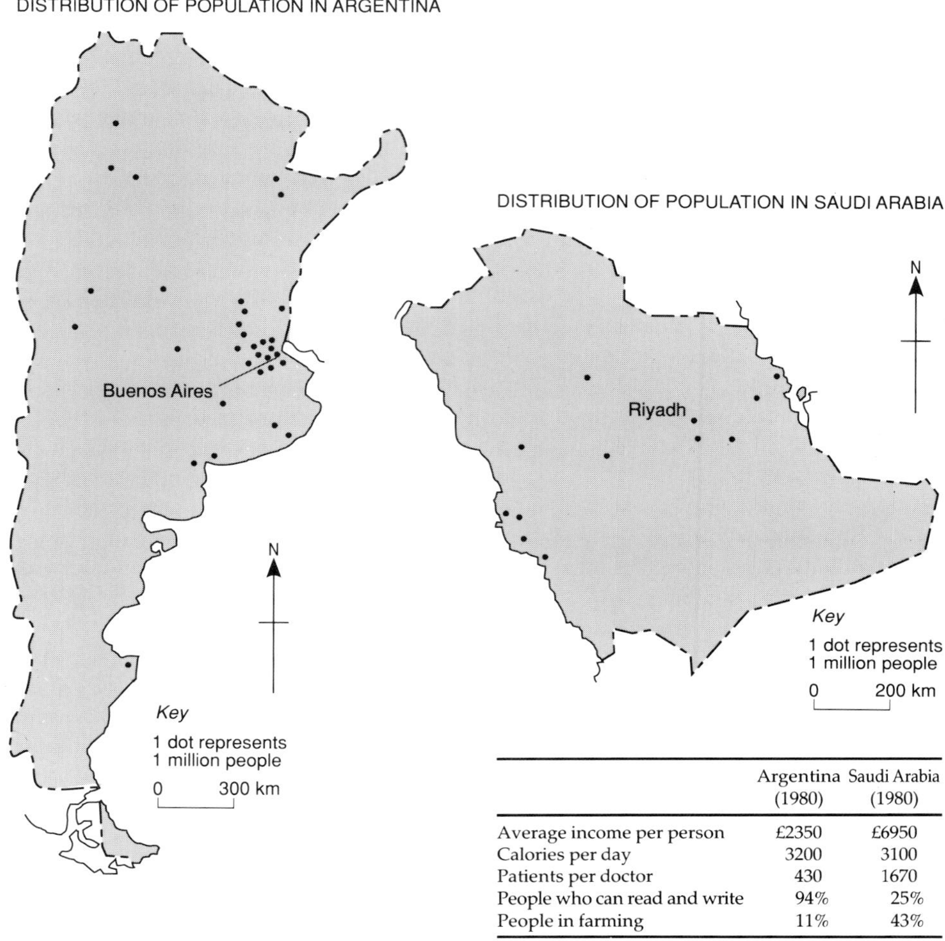

Fig. 2.5

	Population (millions) 1986	Area (millions of sq km)	Population Density
Argentina	31	3	10
Saudi Arabia	12	2	6

Fig. 2.6

	Argentina (1980)	Saudi Arabia (1980)
Average income per person	£2350	£6950
Calories per day	3200	3100
Patients per doctor	430	1670
People who can read and write	94%	25%
People in farming	11%	43%

Fig. 2.7

	Argentina (1960)	Saudi Arabia (1960)
Average income per person	£1700	£2700
Calories per day	2810	1700
Patients per doctor	600	4000
People who can read and write	85%	15%
People in farming	20%	65%

Fig. 2.8

Case Study of the European Community

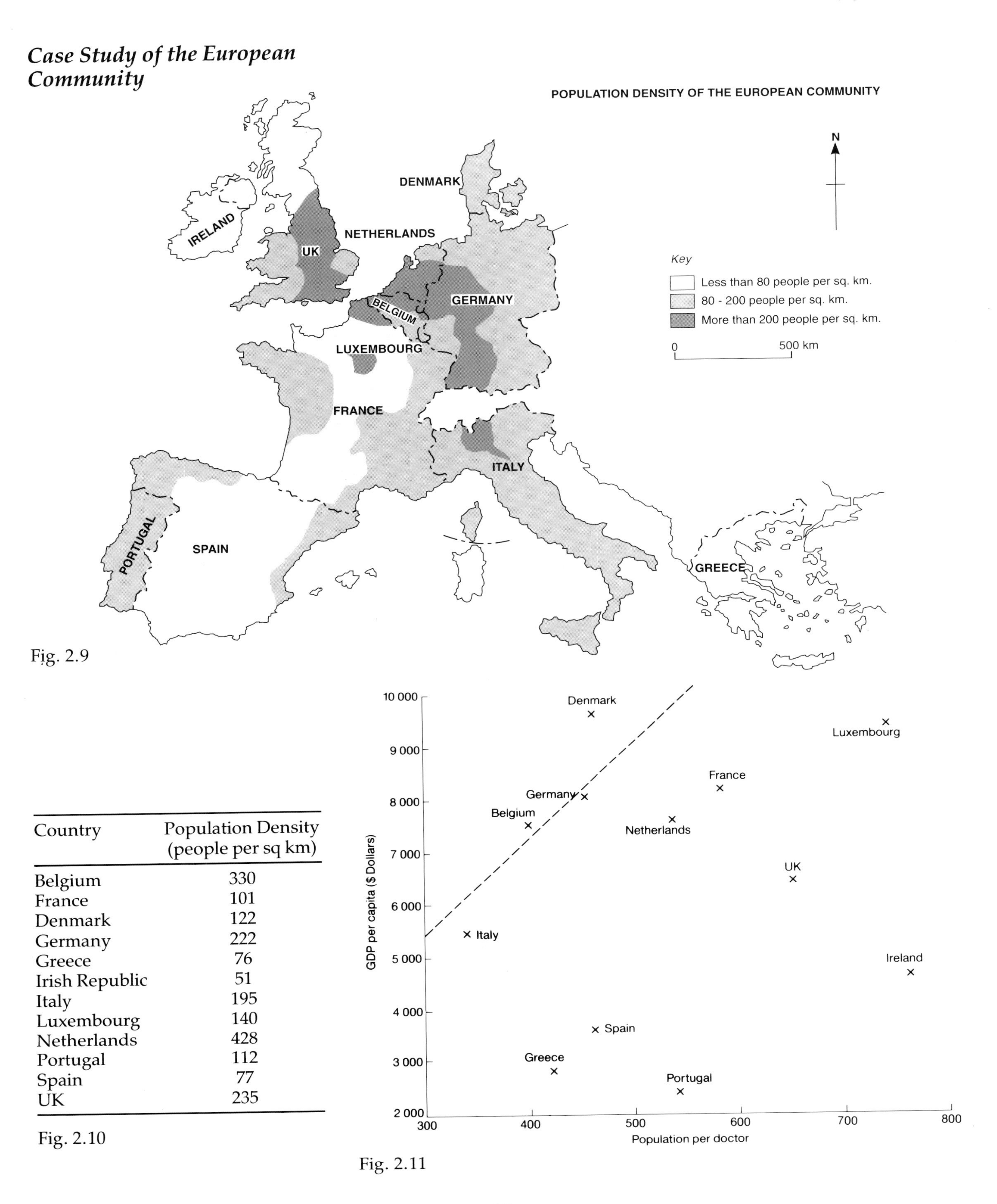

Fig. 2.9

Country	Population Density (people per sq km)
Belgium	330
France	101
Denmark	122
Germany	222
Greece	76
Irish Republic	51
Italy	195
Luxembourg	140
Netherlands	428
Portugal	112
Spain	77
UK	235

Fig. 2.10

Fig. 2.11

Extension Text

2F Indicators of Standard of Living

1 Wealth

A wealthy country can afford to feed its people well and can build schools, hospitals and provide other services so that everyone has a high standard of living.

Indicators of wealth are **average income per person, Gross Domestic Product (GDP) per capita and Gross National Product (GNP) per capita.** GDP per capita is the value of goods and services produced in a country in a year, divided by its population. GNP = GDP + the value of services earned abroad.

2 Food Intake

To have a high standard of living, people must at least have enough food to eat. Many people eat enough calories but do not have a balanced diet. So, as well as **number of calories per person per day**, the **amount of protein per person per day** is often used as an indicator of being well fed.

3 Education

Well educated people can do skilled work, can think of new ideas, can become teachers, doctors, engineers or have other occupations which help to improve everyone's standard of living. The **percentage of people who are literate** does not always tell us exactly how well educated the people are. The **percentage of children at secondary school** is sometimes used instead.

4 Health

People cannot have a high standard of living if they are not healthy most of the time. The **population per doctor** does not tell us how many people are ill. Other useful indicators are **life expectancy** (how long the average person can expect to live) and **infant mortality** (the proportion of children who die before they are 1 year old).

5 Industrialization

If most people work in factories and offices, many goods can be produced and sold abroad. Then the country can afford to import goods which will improve their standard of living.

The **percentage of people in agriculture** does not tell us how many goods are produced. The **amount of energy used per capita** is sometimes used instead.

2G Problems with Indicators

1 If the population has not been counted accurately, all the averages used are unreliable.
2 Averages can hide big differences. Some people may be well fed or wealthy while others may be very hungry or very poor.
3 One indicator of standard of living is not enough. Being able to read does not make up for being hungry. Being wealthy does not make up for being ill.

2H Choropleth Maps

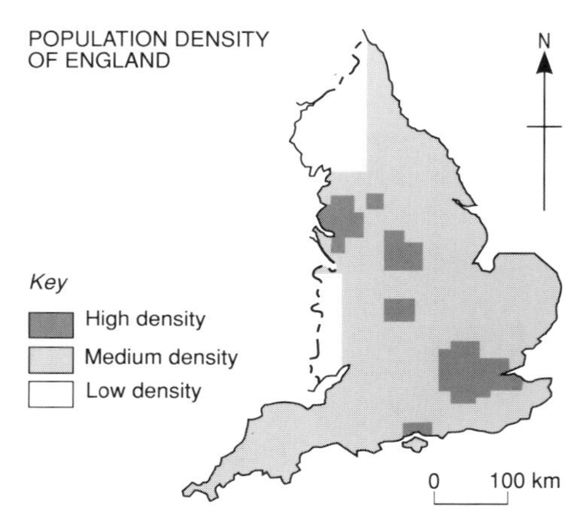

The map above is a **choropleth map** of England's population density. It has been drawn from the map in 2E, by the following method:
1 The map was divided into squares of equal area.
2 The number of people in each area was counted.
3 Each square was shaded according to the population density.

Questions

Look at 2F, 2G and 2H.

E1 What is meant by infant mortality?
E2 What is meant by life expectancy?
E3 What is meant by 'gross domestic product'?
E4 Why is education a good indicator of standard of living?
E5 Why can averages sometimes be misleading?
E6 Why is it better to use more than one indicator to measure standard of living?
E7 Draw a choropleth map to show the population density of Liberia from fig 2.1.

Questions

Case Study of the European Community

'In terms of population density, the European community has a densely populated ***core*** *and a more sparsely populated* ***periphery****.'*
C1 Do you agree with the description above? Give detailed reasons for your answer.

Look at figs. 2.9 and 2.10.

C2 Which shows the more useful way of comparing the population density of the European Community countries, fig. 2.9 or 2.10? Give reasons for your answer.
C3 Describe what the scatter graph, fig. 2.11, shows.

'The area to the left of the dashed line shows the countries with the highest standard of living.'
C4 Do you agree with the statement above? Give reasons for your answer.

UNIT 3 The Empty Lands

Core Text

3A The Empty Lands

Some parts of the world have a high population density. They are called **crowded lands** or **positive areas**. Many parts of the world have a low population density. They are called **empty lands** or **negative areas**. The map opposite shows the main empty lands of the world.

GREENLAND
NEPAL
NEW GUINEA
BOTSWANA

Key
Areas with less than one person per sq. km.
Polar lands
Mountains
Deserts
Tropical rainforests

3B Changes in the Empty Lands

More people now live in the empty lands. This is because
(*a*) **Better technology** (machines, equipment) allows people to cut down forests, bring water to deserts, build tunnels through mountains more easily than in the past. Some areas have been developed for cattle ranching, forestry and tourism.

(*b*) **Minerals** have been found in the empty lands and so many mines and wells have been developed.

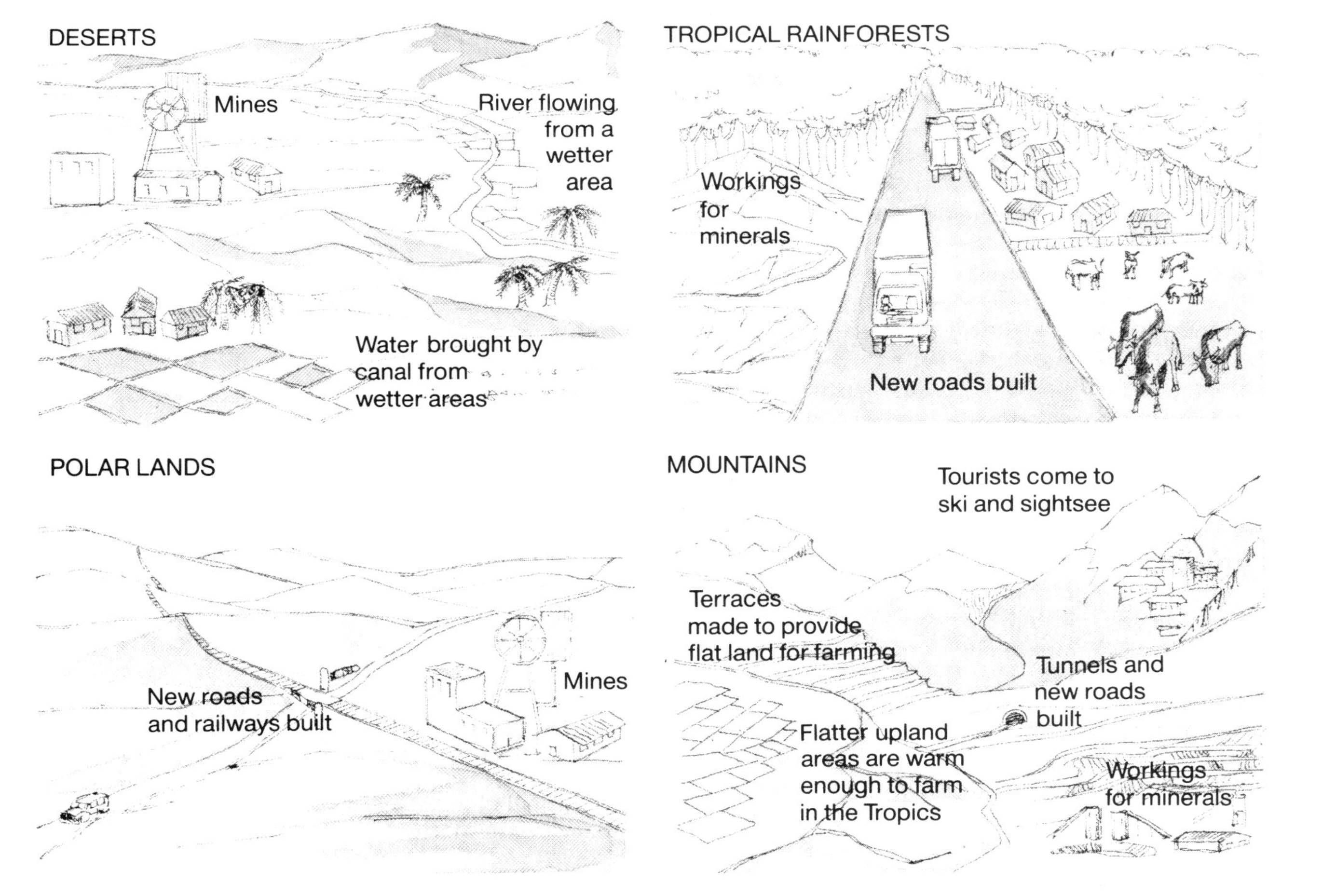

Core Questions

Look at 3A.

1 What are empty lands?

2 Why do few people live in the following countries:
(*a*) Greenland (*b*) New Guinea
(*c*) Botswana (*d*) Nepal?

3 Give two reasons why few people live in
(*a*) polar lands (*b*) mountains
(*c*) rainforests (*d*) hot deserts.

Look at 3B.

4 Explain why there are now more people living in:
(*a*) polar lands (*b*) mountains
(*c*) rainforests (*d*) hot deserts.

5 Why do more people now live in this area?

The hot desert of the USA is a popular place to live. Millions of people live in the desert states of California, New Mexico and Arizona. While many more visit the area to see attractions such as Las Vegas and the Grand Canyon. The River Colorado provides the water which is used in the many new factories here, as well as to water the farmland.

Look at the text above.

5 Why do so many people live in the American hot desert?

Core Investigation

Choose one of the empty lands on the next four pages for your investigation.
The aim of your investigation is:
(*a*) to explain the low density of population in your chosen area, and
(*b*) to describe and explain how life is changing in your area.

Your investigation report should be in four parts:

Aim
Method
Analysis
Conclusion

The rest of this page shows the steps you should take in your investigation and suggestions for how to present it.

How To Write Up Your Investigation Report

Title

Give your investigation a short title eg *A Study of the Flinders Ranges Desert, Australia.*

Use a whole page for your title and your name.

Contents

Divide your report into chapters. The headings below will help you. A well planned report will earn you high marks.

List the chapters on the contents page and, next to each chapter, write down the page at which it begins.

Aim

Start your report with an introduction. In the introduction explain what you are going to investigate.

Try not to use the word 'I' or 'we' eg *The aim of this investigation is to . . . . . .*

Method

In the next chapter write down how you get your information. Most of it will have come from one of the next four pages. But try and find more information if you can eg you could draw a map of the country you are studying from an atlas. You could use books eg to find out more information about the Eskimos or Aborigines. The more research you do the higher your mark.

Write down the title and author of all the books you use and put them in the bibliography at the end of the report.

Analysis

There are two stages to this.

1 Think of different ways of showing your information – not just writing it, but drawing diagrams, bar graphs, line graphs, pie-graphs, cross-sections and maps. For example you might want to draw a graph to show the climate of your area. If you use a lot of different graphs and diagrams, you will get higher marks.

2 Next, write your report explaining what you said you were going to do in your aim. You should have a lot of sentences which begin 'The reasons for this are ....' or 'This is because ....'

You should write down your information in a logical order and give as many reasons as you can. This will earn you high marks.

Conclusion

Finally, explain briefly what your investigation has found out. It should answer the aim that you wrote down at the start eg 'This investigation has shown that the reasons why this area has a low density of population are ....'

Bibliography

List all the books, atlases and other sources that are used.

Case Study of North Slope Borough, Alaska

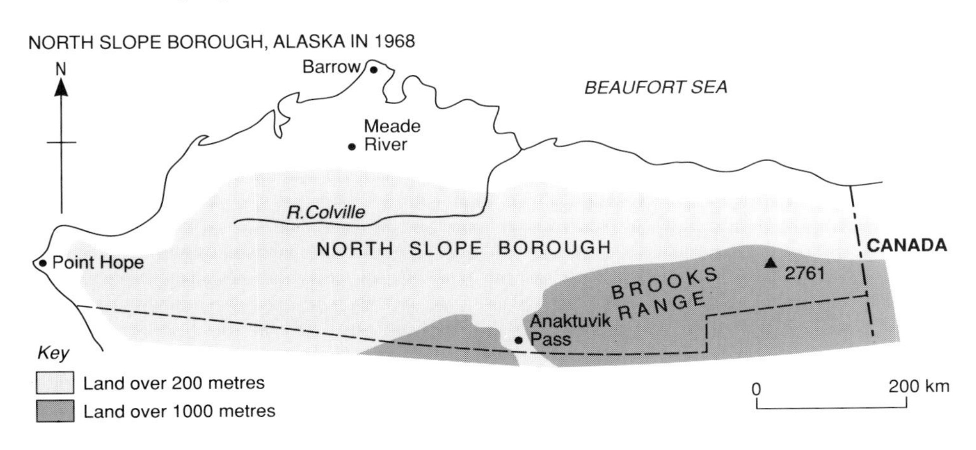

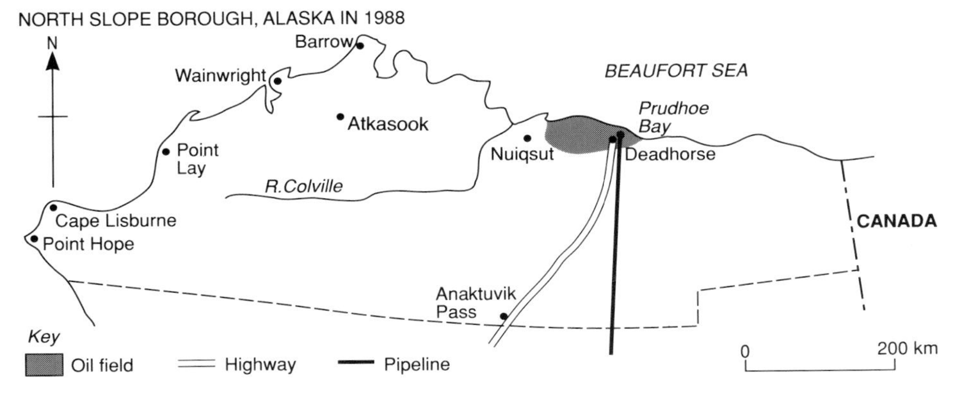

Fig. 3.1

	J	F	M	A	M	J	J	A	S	O	N	D	Total
Precipitation (rain, snow in mm)	10	5	10	5	10	15	20	30	20	15	10	5	155
Temperature (°c)	−28	−30	−25	−16	−5	+3	+7	+6	+1	−9	−19	−25	

22 hours of darkness in the 3 winter months — Temperatures too low for crops to grow — Swarms of flies in summer — Not enough moisture for crops to grow

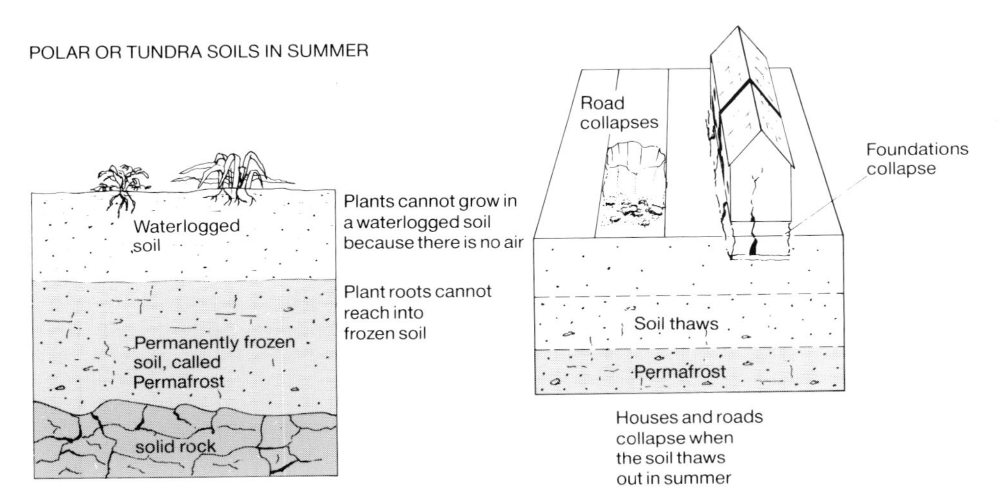

Fig. 3.2

About 4000 Eskimo (or Inuit) people live here. Some still live by hunting caribou, seals and fish. But there are not enough animals for many people to live like this.

Population of North Slope	
1968	3200
1978	4200
1988	5600

Percentage of Eskimo (Inuit) People in North Slope	
1968	88%
1978	77%
1988	65%

Oil was discovered in North Slope Borough in 1968. The United States Government eventually decided that the oil should be extracted. It now produces 2 million barrels per day, worth £20 million, and gives jobs for 5237 people. People have moved here from other parts of the USA to work.

'Frostbite is common in this climate and it is easy to break arms and legs. Tyres stick to the frozen ground. Engines have to be kept running 24 hours a day. White-outs (swirls of wind-whipped snow) make it impossible to find your way.' (Local worker)

Even if oil had been found before 1968 it is only recently that technology has made the building of very long pipelines possible. The need to get oil from such remote areas, at such great expense, is also a recent feature.

The US government will decide what developments take place in Alaska. They might agree with conservationists that this 'last great wilderness' should not be developed any more as this would result in more industry, housing, roads.

Case study of Acre Region, Amazon Rain Forest, Brazil

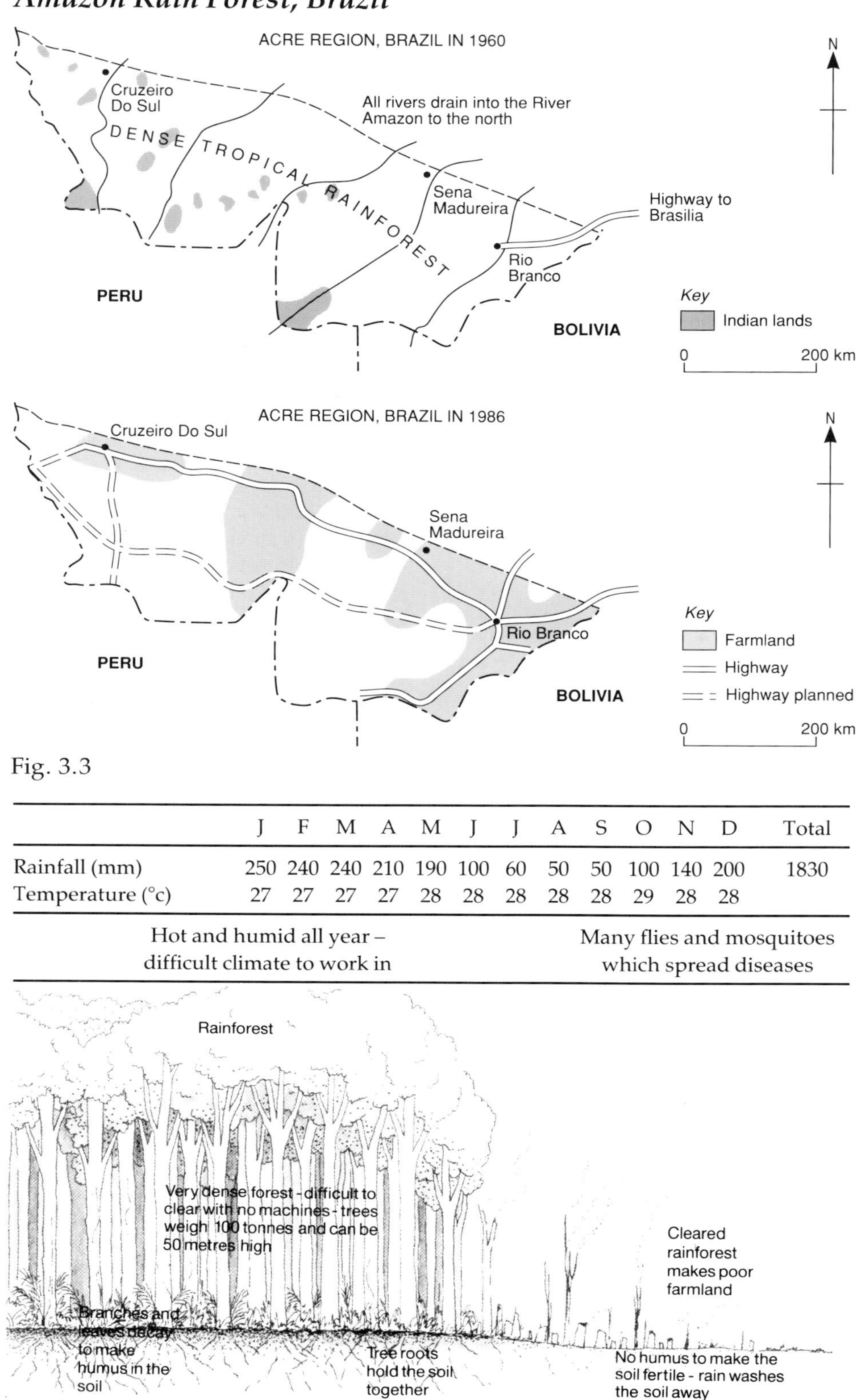

Fig. 3.3

	J	F	M	A	M	J	J	A	S	O	N	D	Total
Rainfall (mm)	250	240	240	210	190	100	60	50	50	100	140	200	1830
Temperature (°c)	27	27	27	27	28	28	28	28	28	29	28	28	

Hot and humid all year – difficult climate to work in

Many flies and mosquitoes which spread diseases

Fig. 3.4

Soil washed away from cleared forest ends in rivers. The rivers rise and flood more often. This makes it difficult to use the land near rivers.

About 9000 Indians live in Acre Region. They clear patches of forest, farm for a few years and then move on when the soil becomes poor. They also hunt animals and fish. But there is not enough space for many people to live like this.

Many people live by collecting rubber and Brazil nuts and selling them in return for food.

Population of Acre Region	
1960	160 000
1970	220 000
1980	300 000

Percentage of Indians in Acre Region	
1960	12%
1970	10%
1980	3%

ROAD DISTANCES FROM RIO BRANCO	
Manaus	1368 km
Brasilia	3033 km
Rio de Janeiro	4024 km
Belem	4857 km
Joao Pessoa	6013 km

New roads through the Amazon Rain Forest to Acre have made it possible for people to keep cattle and grow crops and sell their produce. Many people have moved here from other parts of Brazil.

The Brazilian Government gave money to persuade people to move to Acre Region. It also provided tax incentives for people to set up huge cattle ranches.

It used to take a team of men a whole day to cut down a giant tree with axes. With the arrival of chain saws a single man could do the job in 10 minutes. Now machines can chop an entire tree, branches included, into small chips in 60 seconds.

Aeroplanes are used to seed cleared forest for pasture.

Case Study of the Flinders Ranges Desert, South Australia

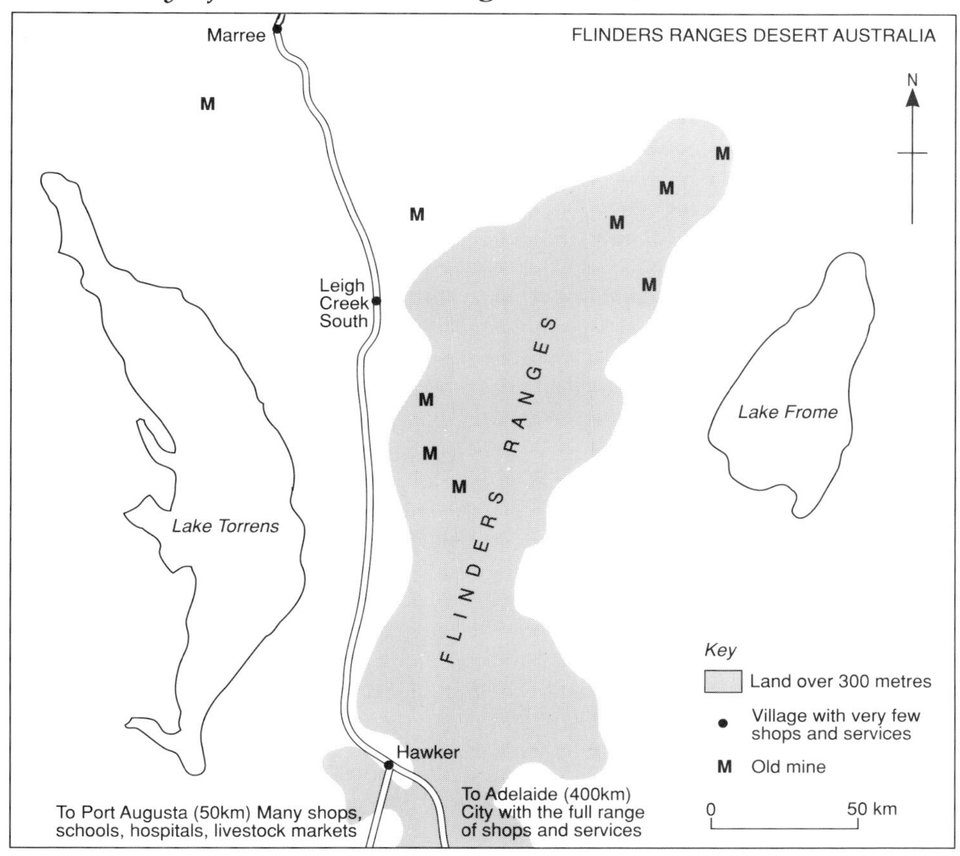

Fig. 3.5

	J	F	M	A	M	J	J	A	S	O	N	D	Total
Rainfall (mm)	10	10	20	20	35	30	25	30	15	10	10	5	220
Temperature (°c)	28	29	28	25	21	19	18	19	21	22	25	27	

Water evaporates quickly in these temperatures

Rainfall is too low to grow crops. The vegetation is not good enough for sheep or cattle

'In summer temperatures can rise to 45°C in the day. Plants are very dry. Fires start easily and, with the strong winds, can travel at over 20 km/hr. They also change direction suddenly. And how do we put them out – in a desert?' (Local resident)

Dust storms are a big problem. They can travel at up to 120 km/hr.

About 300 Aborigines live here. Some survive by eating the few animals in the desert and getting water from plants. But there are not enough animals for many people to live like this.

Australia School of the air. Two way radios are now used by children for their lessons, for talking to other children and teachers far away

Population of the Flinders Ranges	
1961	800
1986	3100

Percentage of Aborigines in Flinders Ranges	
1961	20%
1986	10%

Wind pumps have made it possible for people to farm the desert now. The speed of the wind provides the power which pumps up water from underground.

The farms have to be huge to give enough grazing for the sheep. And the whole farm must be fenced to keep out rabbits and dingoes. Some of the fences might be 50 km from the farmhouse.

Putting extra water on the land is called **irrigation**. In some areas small dams are built to hold water.

'Any attempt to bring water in from wetter areas would be extremely costly and most of it would evaporate before it reached the desert' (Local official)

The Flying Doctor Service now makes it possible to get people to hospital quickly.

Without the road, it would not be possible to get wool and sheep to market cheaply and quickly enough to sell it.

Coal was mined at Leigh Creek but the miners were defeated by the area's remoteness and drought.

Case Study of Livigno, Italian Alps

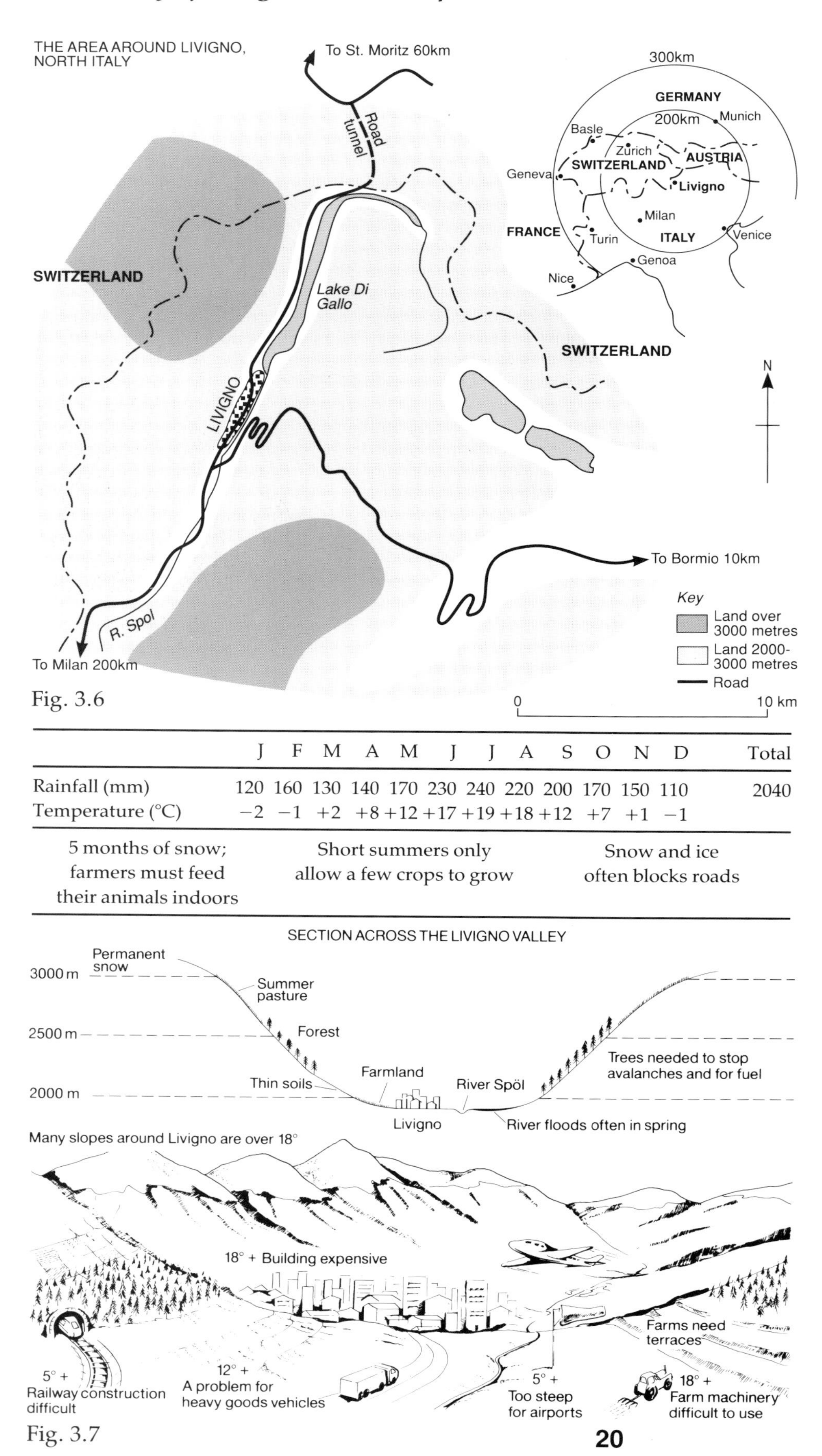

Fig. 3.6

	J	F	M	A	M	J	J	A	S	O	N	D	Total
Rainfall (mm)	120	160	130	140	170	230	240	220	200	170	150	110	2040
Temperature (°C)	−2	−1	+2	+8	+12	+17	+19	+18	+12	+7	+1	−1	

5 months of snow; farmers must feed their animals indoors	Short summers only allow a few crops to grow	Snow and ice often blocks roads

Fig. 3.7

In 1961, most people in Livigno were farmers, keeping sheep and cattle. In summer they grew hay and a few crops in the valley, while their cattle were taken to the high Alpine pastures. This is called **transhumance farming**. The land was not good enough for many people to make a living from farming. Many young people moved away to find work in the cities.

Livigno	1961	1986
Population	1700	3900
People in farming	40%	25%
People in Tourism	1%	10%

In 1970, a 3 kilometre road tunnel was built through the mountains near Livigno. This made it much easier for people to reach this area. Now Livigno's 5 months of snow make it ideal for skiers. Many people now have jobs in hotels, shops and restaurants.

EXTRACT FROM A SKIING BROCHURE

There are many off the slopes activities to choose from here. As Livigno is considered the Italian capital for free-style skiing, there are often exciting displays to watch and the resort also boasts a horseriding centre and 2 natural ice rinks. St.Moritz is just a few hours drive away over the Swiss border and excursions are arranged every week. Bormio and Santa Caterina can also be easily reached on the local bus or if you prefer to stay in Livigno why not take a romantic horse drawn sleigh ride? There are over 100 shops in Livigno selling everything from fur coats and computers to tobacco and chocolate, not to mention the wines and spirits at duty-free prices! Finally if you are thinking of investing in some ski equipment, it is also very reasonably priced here.

Fig. 3.8

Skiing is one of Europe's most popular sports and it is becoming more and more popular. More people can now afford to take a winter holiday and many cheap 'package holidays' are available.

Livigno is a duty-free resort because, in the past, it was too remote for officials to come and collect the duties. Now its duty-free status means lower prices, more shops and more tourists.

UNIT 4 The Crowded Lands

Core Text

4A The Crowded Lands

The **crowded lands** are those areas in the world which have a high population density. The map below shows where the main crowded lands are found.

CROWDED LANDS OF THE WORLD

North-west Europe
Ganges valley, India
East China
East Japan
North-east U. S. A.
Nile valley, Egypt
South India
Java
South-east Brazil

Key
Areas with more than 200 people per sq. km.

There are many reasons why some areas are crowded. The most important reasons are to do with the environment. They are called **environmental** or **physical factors.**

Housing in Kowloon

4B Reasons Why Areas Are Crowded – Environmental Factors

1 LOWLANDS WITH GENTLE SLOPES

Most people live in lowlands and on gentle slopes. This makes it easy to build houses, factories, roads, and railways. It is also easier to use farm machines on gentle slopes.

2 MINERAL RESOURCES

Areas with valuable mineral resources are often crowded. For example, many people live on coalfields. This is because there are jobs in mining, but also because there are many factories on coalfields which also provide work. Areas with oil, iron ore and other valuable minerals can also have a high population density.

3 WATER SUPPLY

A reliable supply of water – from rivers, reservoirs or wells – encourages people to live in an area. They have enough water for themselves, their factories and their crops.

4 SUITABLE CLIMATE

Most people live in areas which are not too hot nor too cold; and in areas which are not too wet nor too dry. In these climates it is much easier to farm and it is also more pleasant to live.

5 FERTILE SOILS

Areas with deep fertile soils are popular. In these areas people can grow a lot of food from just a small area of farmland.

4C Examples of Crowded Lands

The table below shows why some areas of the world are so crowded.

	Java	North-east USA	Nile Valley, Egypt	Ganges Valley, India	South India	South-east Brazil	East China	East Japan	North-west Europe
Lowland		✓	✓	✓	✓		✓	✓	✓
Suitable Climate	✓	✓		✓	✓	✓	✓	✓	✓
Fertile Soil	✓		✓	✓	✓	✓	✓		
Mineral Resources		✓				✓	✓		✓
Water Supply	✓	✓	✓	✓	✓	✓	✓	✓	✓

Core Questions

Look at 4A.

1 Name five crowded lands in the world.

2 Give one word for the population density of the crowded lands.

Look at 4B.

3 Why do many people live on gentle slopes?

4 Why do many people live on fertile soils?

5 Why do many people live on coalfields?

Look at 4C.

New York

6 Why is North-east USA so crowded?

7 Why does Java have a high population density?

8 Why does the Nile Valley have a high population density?

9 Why is the Ganges Valley so crowded?

10 Which crowded areas are not in lowlands?

Questions

Case study of Scotland

Look at fig. 4.9.

F1 Which is (a) the most crowded and (b) the least crowded part of Scotland?

'The Central Lowlands is crowded because it is much flatter than the rest of the country.'

F2 The statement above is not completely true. Explain why.

'People in Scotland live where there are natural resources.'

F3 Do you agree with the statement above?
Give a reason for your answer.

	Population in 1961 (thousands)	Population in 1985 (thousands)
Central Lowlands	4697	4607
Highlands	234	278
Southern Uplands	248	245

Look at the table above.

F4 Which areas of Scotland have become more crowded since 1961?

'Nothing can be done to make the Highlands more attractive for people to live in.'

Look at the diagram above.

F5 Explain how the statement above is exaggerated.

Questions

Case Study of China

Look at fig. 4.1.

G1 Describe the population density in China.

	China	North-east	North-west	South-east	South-west
Lowland					
Suitable climate					
Fertile soil					
Mineral resources					
Water supply					

G2 Copy the table above and use Fig. 4.1 to complete it.

'North-east and South-east China are crowded because they have a lot of mineral resources.'

G3 Do you agree with the statement above? Give reasons for your answer.

'North-west China has nothing to attract people to live there'

G4 Give one argument for and one argument against the point of view above.

Look at fig. 4.2.

G5 Describe changes in the population density of China between 1953 and 1986.

'The changes in population density in China since 1953 have been mostly due to Government measures.'

Look at figs. 4.2 and 4.3.

G6 Give one argument for and one argument against the point of view above.

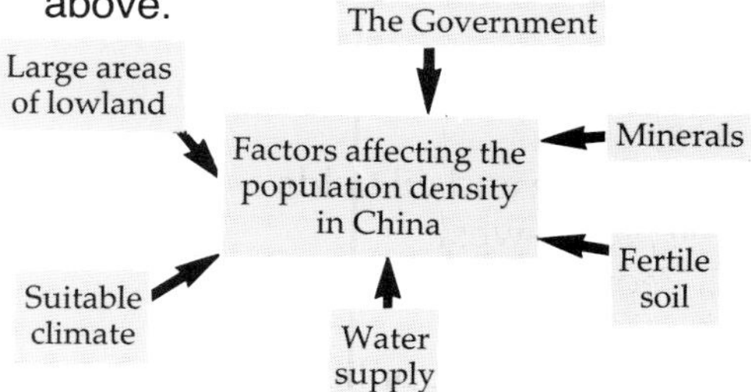

G7 Describe in detail what the flow chart above shows?

Resources

Case Study of China

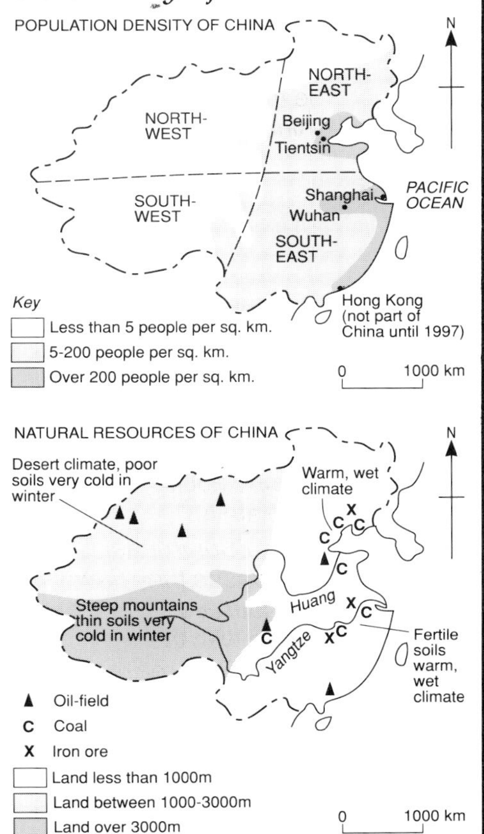

Fig. 4.1

Region of China	Population Density (people per sq km) 1953	1986
North-east	50	100
South-east	174	324
South-west	40	71
North-west	11	22

Fig. 4.2

Some Government Measures Since 1953

1 Iron-ore and oil fields opened in the North-west
2 People moved out of the cities in the East
3 New water supplies developed in the South-West

Fig. 4.3

Case Study of the Netherlands

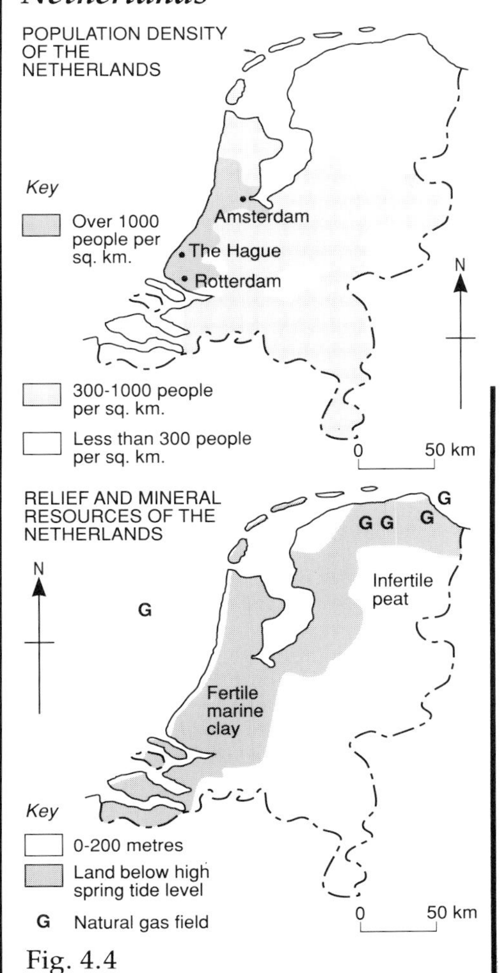

Fig. 4.4

The most fertile soils are in the areas below sea level. But only 6 per cent of Dutch people work in agriculture. The climate does not vary much across the country.

Fig. 4.5

In its natural state, nearly half of the Netherlands would be salt-marsh, lakes and bog. But with great skill, the Dutch have reclaimed this land and turned it into highly productive farmland.

Fig. 4.6

The Dutch Government has tried to reduce the population density of its crowded areas by giving grants to factories which set up in other parts of the country.

Fig. 4.7

Region of Netherlands	Population Density (people per sq km) 1977	Percentage Change in Density 1977-1988
Northern	164	+12%
Eastern	311	+ 7%
Central	901	+ 5%
Southern	380	+ 2%

Fig. 4.8

Case study of Scotland

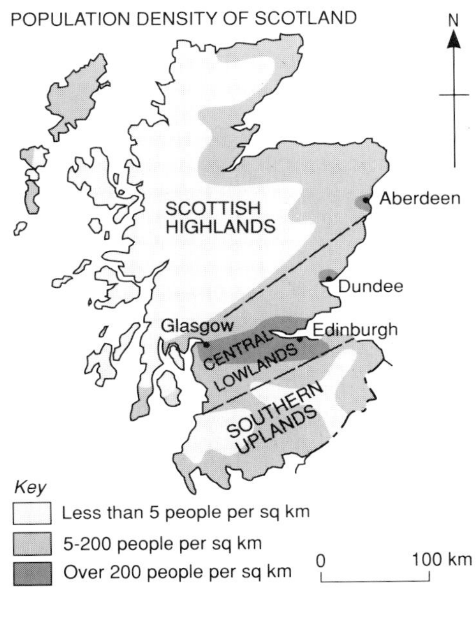

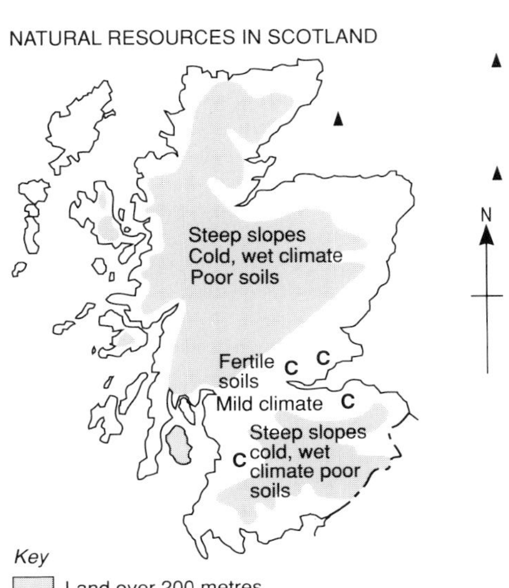

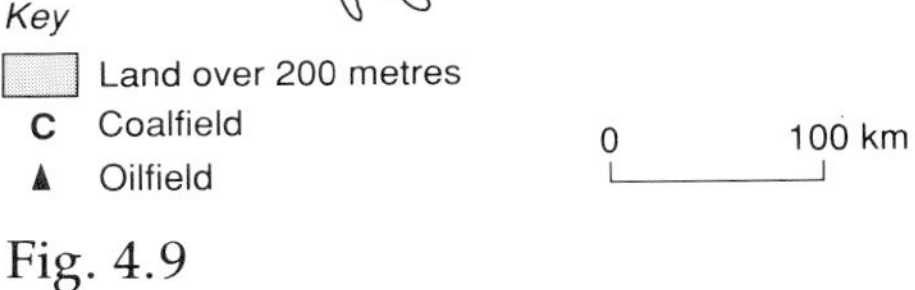

Fig. 4.9

Extension Text

4D Reasons Why Areas are Crowded – Human Factors

On a world scale, the density of population is affected mostly by physical or environmental factors eg relief, climate, soil. On a regional scale, environmental factors are still important but so are **human factors**. People have the ability to make the natural environment better. Then more people can live there.

1 Economic Factors

Economic factors are those which concern the wealth and development of a country. A wealthy country can build many roads, railways, airports. These attract industries which, in turn, encourage a high population density.

A wealthy country can also use its technology to overcome problems of relief (eg by tunnels, bridges), climate (eg by irrigation) and soil (eg by fertilisers). Technology makes it possible for a lot of people to live in an area which otherwise would be sparsely populated.

If the people in a country are skilled and educated eg Japan or Sweden, they will be able to find more employment eg by inventing products, working efficiently, providing specialist services. With a lot of employment, that area is likely to have a high population density.

2 Political Factors

Political factors are those which concern the government and the decision it makes. If the government of a country decides to invest a lot of money and technology in one region eg building roads or irrigation schemes, more people will go and live there and the population density will rise. The opposite is also true.

4E Summary

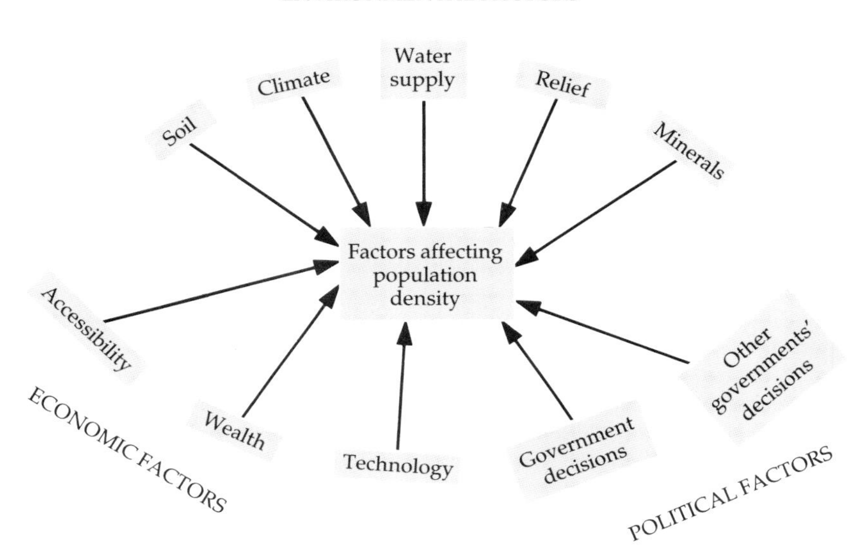

Questions

Look at 4D.

E1 Which factors most affect population density on a world scale?

E2 Which factors most affect population density on a regional scale?

E3 How can the wealth of a country affect its population density?

E4 Explain how greater technology can permit a high density of population.

E5 How can a skilled and educated people make possible a high population density?

E6 Explain how governments can influence population density.

Questions

Look at figs. 4.4 and 4.5.

C1 Compare the natural environment of the most crowded and least crowded parts of the Netherlands.

C2 Do you think that variations in population density in the Netherlands can be explained by environmental factors? Give detailed reasons for your answer.

'Government attempts to spread the population more evenly throughout the Netherlands have failed.'

Look at figs. 4.7 and 4.8.

C3 Explain how the statement above is exaggerated.

C4 Which are most important in explaining the population density of the Netherlands – environmental, economic or political factors?

C5 Draw a flow chart (similar to 4E) to show the factors affecting the population density of the Netherlands. More important factors should have thicker arrows than less important factors.

UNIT 5 Population Growth

Core Text

5A *The World's Population*

The population of the world is growing very quickly. The table below shows world population growth over the last 300 years. It is growing because there are more births in the world than deaths. To find out why the population of the world is growing, we must find out why there are more births than deaths.

Year	World population (millions)
1700	600
1750	700
1800	900
1850	1100
1900	1600
1950	2300
2000	6000

5B *Birth Rates and Death Rates*

Birth Rate = the number of births for every 1000 people each year.

Each year there are 28 births for every 1000 people in the world.

This can be written as 28‰.

Death Rate = the number of deaths for every 1000 people each year.

Each year there are 10 deaths for every 1000 people in the world.

This can be written as 10‰.

The Natural Increase in Population (the number of extra people) = **Birth Rate − Death Rate**

The natural increase in the world = 28‰ − 10‰ = 18‰.

So for every 1000 people in the world there are 18 more at the end of the year.

5C *Differences in Birth Rates*

The average birth rate in the world is 18‰, but some countries have a higher birth rate and some lower.

Country	Birth Rate ‰
Argentina	21
India	35
Ethiopia	49
Liberia	50
Netherlands	13
United Kingdom	12

DEVELOPED COUNTRIES = LOW BIRTH RATE
DEVELOPING COUNTRIES = HIGH BIRTH RATE

In many parts of the Developing World the birth rate is higher than in Developed Countries. There are several reasons for this:

(a) people have large families to work on the land and go out to work,
(b) they need children to look after old or sick parents,
(c) family planning is not generally approved of because of the high death rate among babies.

Whereas in the Developed World;

(a) it is expensive to have a large family,
(b) women may have to give up work to look after their babies, which may reduce the family income,
(c) children can be a 'tie' and a responsibility, so some couples only have children in their 30s.

5D *Differences in Death Rates*

The average death rate in the world is 10‰, but some countries have a higher death rate and some lower.

Country	Death Rate ‰
Afghanistan	23
Australia	8
Ethiopia	25
Japan	6
Nigeria	18
United Kingdom	12

DEVELOPED COUNTRIES = LOW DEATH RATE
DEVELOPING COUNTRIES = HIGH DEATH RATE

The cartoon below shows some of the reasons why the death rate is usually much higher in the Developing World.

DEVELOPING WORLD

DEVELOPED WORLD

Core Questions

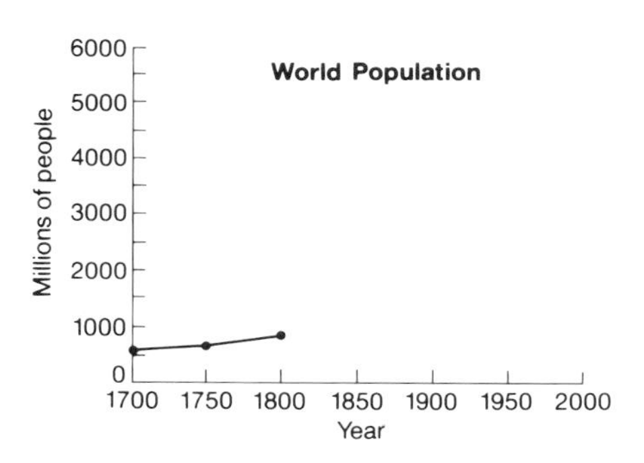

1 Using the population figures in 5A, copy and complete the graph above, which shows the world population figures on the vertical axis and the year on the horizontal axis.

2 ***Look at 5B.***
Spain has a birth rate of 13‰. Does this mean:
(a) 13 people are born each year
(b) for every 1000 people, there are 13 born each year
(c) 1300 people are born each year?

Country	Birth Rate ‰	Death Rate ‰	Natural Increase ‰
Afghanistan	49	23	26
Australia	17	8	
Ethiopia	49	25	
Japan	15	6	
Liberia	50	14	
Italy	13	9	

3 Copy the table above. Complete the table by working out the natural increase in population each year.

Natural Increase = Birth Rate – Death Rate

4 ***Look at 5C.***
Give reasons why people in Developing Countries often have many children.
5 Give reasons why people in Developed Countries often have few children.

6 ***Look at 5D.***
Explain why few people in Developing Countries live until old age.
7 Explain why most people in Developed Countries live until old age.

Questions

Case Study of India

Look at figs. 5.1 and 5.2.
F1 Draw and complete the graph of India's population growth.
F2 How has India's population changed since 1950?
F3 (a) Draw and complete fig. 5.3.
(b) In which year did India's population grow most quickly?

F4 ***Look at fig. 5.4.***
In 1950 there were no old age pensions in India.
Do you think this affected the number of children the people had? Give a reason for your answer.
F5 *'In 1950 people would have had fewer children if there had been more birth control measures.'*
Do you agree with this point of view? Give a reason for your answer.

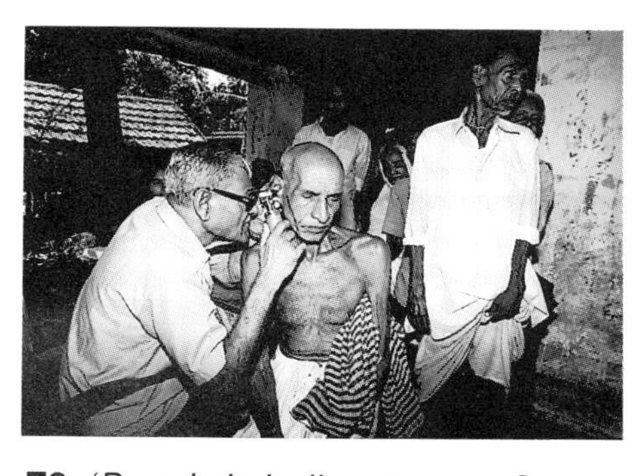

F6 *'People in India are poor. So they should not have many children'.*
This statement is unfair. Explain why.
F7 In 1950, a fifth of all babies born in India died before they were one year old. Do you think this affected the birth rate in India? Give a reason for your answer.

Questions

Case Study of Germany

G1 ***Look at fig. 5.6.***
Draw a graph to show the change in population in Germany since 1950.

G2 ***Look at fig. 5.7.***
Describe how the birth rate in Germany has changed since 1950.
G3 Describe how the natural increase in population in Germany has changed since 1950.
G4 There is compulsory education in Germany until people are 18 years old.
Do you think that this affects the birth rate? Give reasons for your answer.

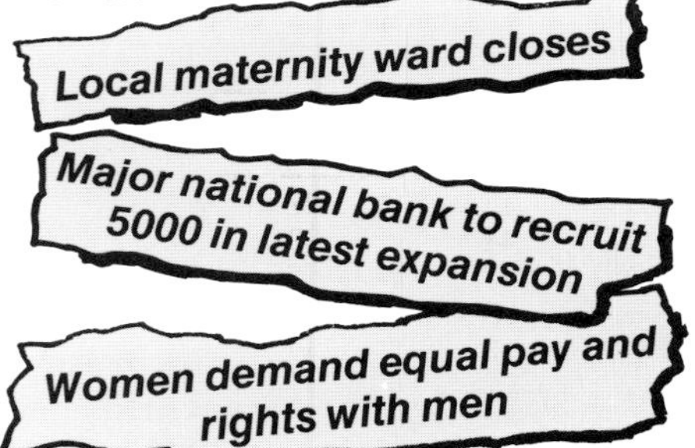

G5 ***Look at fig. 5.8.***
Do you think the trends mentioned above will affect birth rates in Germany? Give reasons for your answer.

G6 ***Look at fig. 5.9.***
Which do you think is more important in explaining Germany's low death rate – (a) free medical care or (b) plenty of food for everyone? Give reasons for your answer.
G7 *'Germany has a low death rate simply because it is a rich country'.*
Give an argument for and an argument against this point of view.

Resources

Case Study of India

Year	Population (millions)
1950	350
1960	430
1970	550
1980	690
1990	870

Fig. 5.1

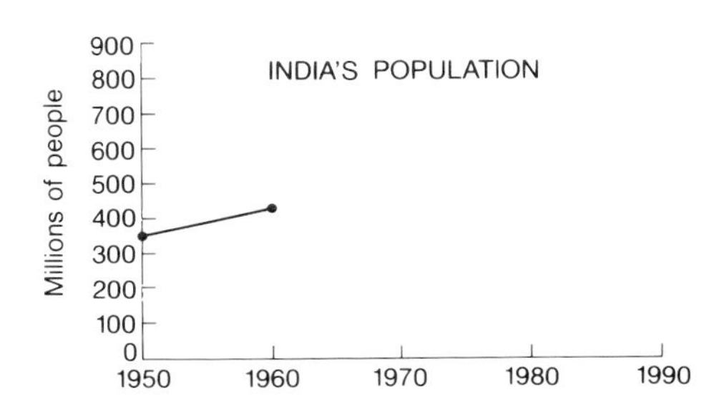

Fig. 5.2

Year	Birth Rate (‰)	Death Rate (‰)	Natural Increase (‰)
1950	42	27	
1960	44	22	
1970	42	19	
1980	35	15	
1990	32	14	

Fig. 5.3

In 1950 75 per cent of Indian families earned a living from farming. There were no old age pensions nor social security payments. There were few birth control measures available. Children did not have to go to school and most couples got married when they were still teenagers.

Fig.5.4

In 1950 India was a very poor country and could not afford to spend much money on medicines and hospital equipment. Each doctor had at least 3000 patients and there were many diseases eg malaria, cholera, smallpox. The water people drank was often polluted and this spread disease. Many people did not get enough to eat.

Fig.5.5

Case Study of Germany

Year	Population (millions)
1950	50
1955	53
1960	55
1965	58
1970	61
1975	62
1980	62
1985	61
1990	61

Fig. 5.6

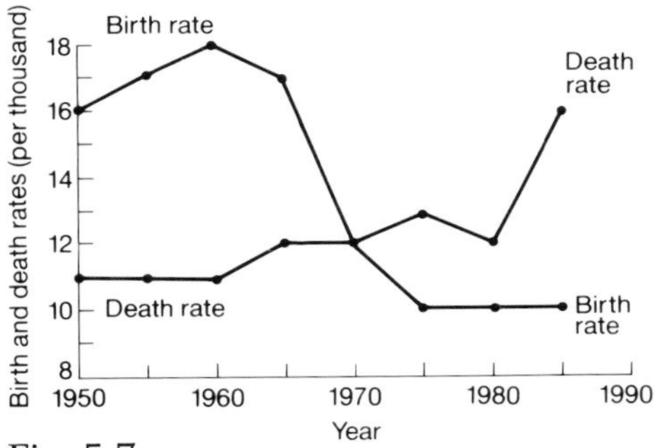

Fig. 5.7

In Germany most people work in factories and offices. The country can afford old age pensions and social security payments. Children must stay at school until they are 15 and then have vocational education until they are 18. Birth control measures are widely available. Many women have a career and couples often do not get married until they are in their mid twenties.

Fig. 5.8

Germany can afford the best medical equipment and hospitals. It spends a lot of its wealth in providing free medical care and free immunization. Everyone has enough food to eat and they drink water which has been purified. Everyone can read so it is easy to educate people about health matters. There are only 450 people per doctor, one of the lowest ratios in the world.

Fig. 5.9

Case Study of Egypt

Year	Birth Rate (‰)	Death Rate (‰)	Population (millions)
1800	47	44	2
1830	45	42	3
1860	46	38	6
1890	48	35	9
1920	46	29	13
1950	45	22	20
1980	38	13	42

Fig. 5.10

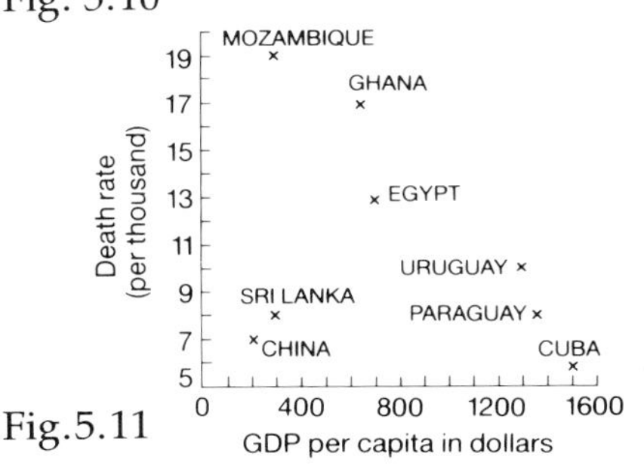

Fig.5.11

Factors in High Birth Rate (1950)	
% in agriculture	57%
GDP per capita	£200
% children at secondary school	10%
Average age of marriage	18
Family planning clinics	nil
Infant mortality rate	190‰

Fig. 5.12

Factors in High Death Rate (1950)	
GDP per capita	£200
Population per doctor	1500
People with safe water	40%
Calories per person	2360 per day
Literacy rate	25%

Fig. 5.13

Extension Text

5E Model of Population Change

Every country's birth and death rates change over time. The graph below shows how they usually change. Because it shows how the birth and death rates of most countries change, it is called a **model of population change.**

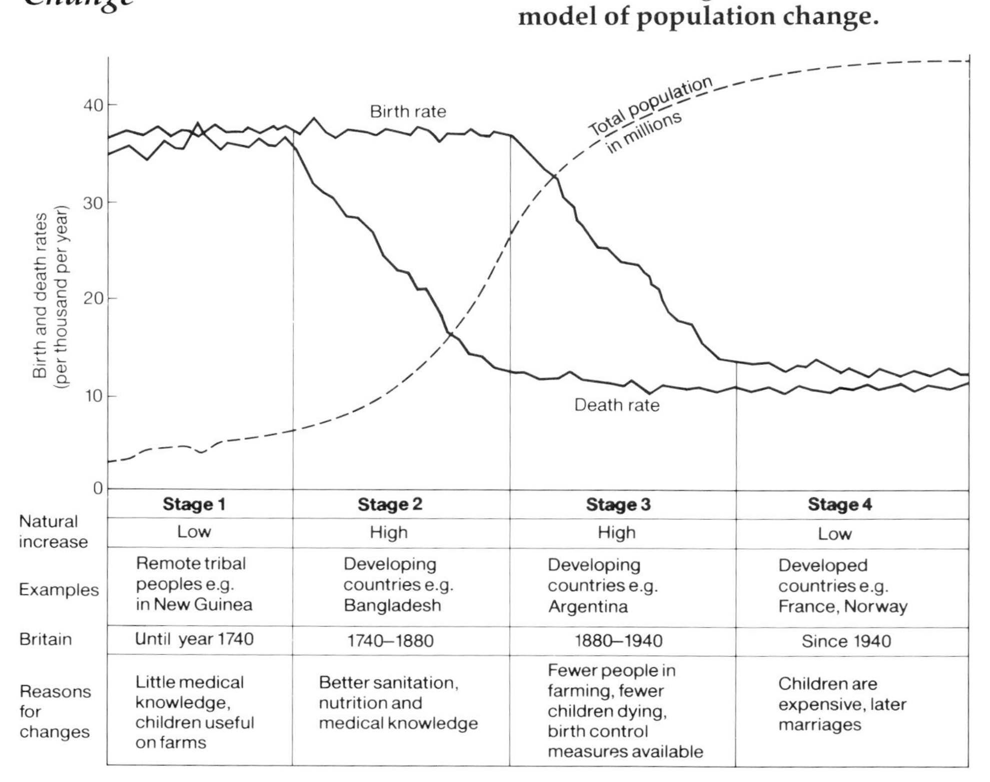

	Stage 1	Stage 2	Stage 3	Stage 4
Natural increase	Low	High	High	Low
Examples	Remote tribal peoples e.g. in New Guinea	Developing countries e.g. Bangladesh	Developing countries e.g. Argentina	Developed countries e.g. France, Norway
Britain	Until year 1740	1740–1880	1880–1940	Since 1940
Reasons for changes	Little medical knowledge, children useful on farms	Better sanitation, nutrition and medical knowledge	Fewer people in farming, fewer children dying, birth control measures available	Children are expensive, later marriages

5F Other Measurements of Birth Rate

The **Crude Birth Rate** is the number of children born for every 1000 people. This can be misleading. It depends how many young women there are in the population. It is better to work out the number of children born in relation to the number of women of child bearing age. This is called the **General Fertility Rate.**

$$\text{General Fertility Rate} = \frac{\text{Total births in 1 year}}{\text{Number of women aged 15-49}} \times 1000$$

5G Other Measurements of Death Rate

The **Crude Death Rate** is the number of deaths for every 1000 people. This can also be misleading. It depends how many old people there are in the population. It is better to work out the number of deaths in each age group in relation to the number of people in that age group. This is called the **Age Specific Death Rate.**

$$\text{Age Specific Death Rate for over 60s} = \frac{\text{Number of deaths among over 60s}}{\text{Number of people over 60}} \times 1000$$

Also used is the **Infant Mortality Rate**

$$\text{Infant Mortality Rate} = \frac{\text{Number of infant deaths}}{\text{Number of live births}} \times 1000$$

Questions

Look at 5E.

E1 What is meant by a 'model of population change'?

E2 Explain why there is a high birth rate in stage 1 of the model.

E3 Explain why there is a lower death rate in stage 2.

E4 Explain why there is a lower birth rate in stage 3.

E5 Explain why there is a low birth rate in stage 4.

E6 Copy the table below.

Country	Birth Rate (‰)	Death Rate (‰)	Stage
Sweden	12	11	
Cambodia	31	29	
Nigeria	50	18	
China	21	7	

Use 5E to complete it.

E7 ***Look at 5G.***
Why can the crude death rate sometimes be misleading?

Questions

Case Study of Egypt

Look at fig. 5.10.

C1 Draw a composite line graph to show the birth rate, death rate and population of Egypt since 1800. Use two separate scales one for the population and one for the birth and death rates.

C2 Describe the relationship between birth rate, death rate and population change since 1800.

C3 Compare changes in Egypt's birth and death rates with the model of population change (5E).

C4 Rank the factors in fig. 5.12 according to how important they are in explaining Egypt's high birth rate. Explain the ranking that you have made.

Look at figs. 5.11 and 5.13.

C5 *'Egypt's high death rate is entirely due to lack of money'*
Explain the different points of view that people would have towards the statement above.

Core Groupwork

China has a quarter of all the people in the world – over 1000 million. Everyone in China agrees that the increase in population must slow down. But the people cannot agree on how this should be done.

Task 1 In groups of four decide which person should be
(*a*) the People's Representative for the Towns
(*b*) the People's Representative for the Countryside
(*c*) the Government Minister
(*d*) the Head of Government.

Task 2 Each person reads the information which relates to them and then makes decisions on the questions at the end of the page.

Task 3 The two Peoples' Representatives and the Government Minister give their decisions to the Head of Government. They must give good reasons for their decisions.

Task 4 The Head of Government listens to the three views and then decides what China's population policy will be.

Task 5 The Head of Government is spokesperson for that group and reports to the rest of the class.

People's Representatives for the Countryside

1 Farmers find that children are useful because they can help on farms.
2 Many farmers are quite wealthy.

People's Representative for the Towns

1 Most people in towns live in small apartments with small gardens or no gardens at all.
2 Most people in towns are quite poor.

3 It is the custom for males to look after their parents when the parents are too old to work.
4 When couples were only allowed one child tens of thousands of baby girls were murdered, so that couples could try again to have a boy.
5 Children have to go to school for nine years and parents have to pay towards the cost.
6 A Chinese woman wielding an axe was reported to have shouted at a birth control official. *'If you sterilize me, I will kill all of you.'* (Daily Telegraph)

Statements 3-6 refer to both representatives.

Head of Government

1 You want to keep the people in the towns and countryside happy. You also want everyone to be well fed and for the country to develop.
2 You must listen to the three views and base your decisions on these views. You could make a compromise if necessary eg allow a higher number of children if they are born five years apart.
3 Three quarters of the people live in the countryside.

Government Minister

1 You want the country to be able to grow enough food for all your people. You also want to develop the country. Old age pensions cost money. So do extra police and courts, which will be needed if the population laws are unpopular.
2 20 million people are short of food now.
3 One child per couple means 15 million extra people in 12 years. Two children means 20 million extra. Three children means 25 million extra, and four children means 30 million extra.
4 There are 50 million old people in China.

Questions

1 How many children should couples be allowed?
Either 1, 2, 3 or 4.

2 When should couples be allowed to marry?
Either aged 18, 22, 26 or 30.

3 How should the government enforce the law?
(*a*) Heavy fines for couples that break the law.
(*b*) Benefits if couples keep within the law eg free pensions when they are old.
(*c*) Sterilize couples when they have had the permitted number of children.

UNIT 6 The Effects of Population Change

Core Text

6A Differences in Population Growth

The population of the world is growing rapidly. Some countries are growing at a very rapid rate. Others are hardly growing at all. The natural increase in population tells us the number of extra people in a country's population each year.

Country	Birth Rate ‰	Death Rate ‰	Natural Increase ‰
Denmark	13	11	2
Sweden	12	11	1
USA	16	9	7
Iran	44	14	30
Kenya	56	14	42
Vietnam	40	14	26

DEVELOPED COUNTRIES = LOW NATURAL INCREASE

DEVELOPING COUNTRIES = HIGH NATURAL INCREASE

6B Effects of a Rapidly Growing Population

1 A lot of children and, in the future, a lot of adults.

2 Problems of feeding all the extra people each year.

3 Not enough houses and services for everyone.

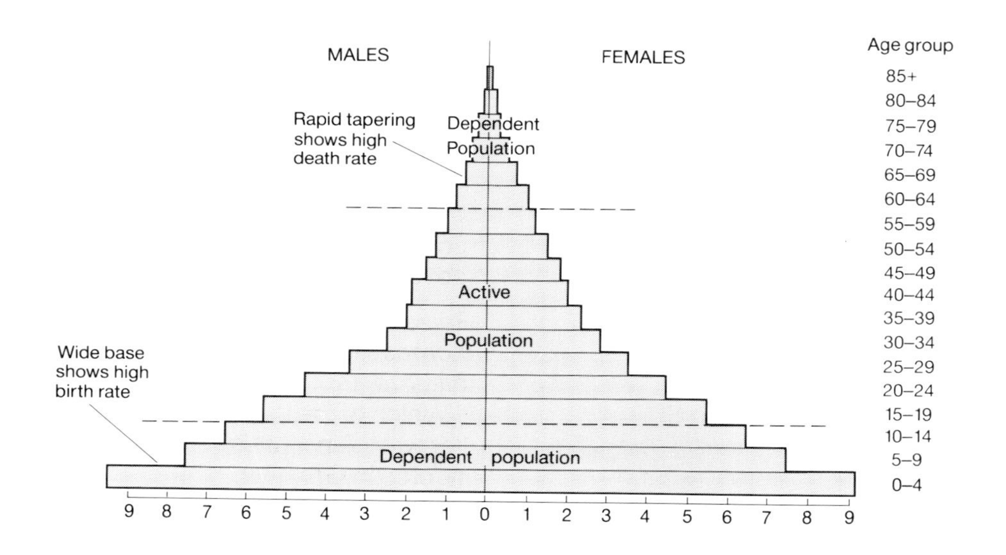

6C Effects of a Slowly Growing Population

1 A lot of old people and few young children.
2 In future, few working adults.
3 A lot of working adults at present.

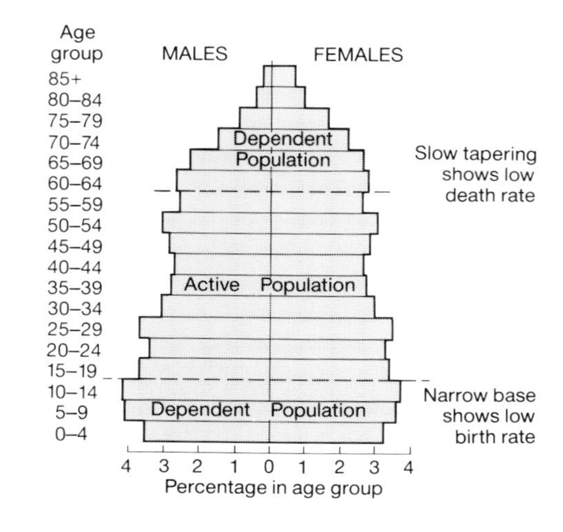

6D Solving the Problems of a Rapidly Growing Population

Solution 1 – Reduce Birth Rate

1 Give advice on birth control at family planning clinics. Train people to give advice.
2 Fine people who have large families.
3 Reward couples who only have one child, eg give free education.
4 Sterilize couples.
5 Not allow people to marry until they are 25 years old.

Solution 2 – Develop The Country's Wealth

1 Find more resources eg coal, oil, iron ore.
2 Use farmland better eg fertilizers, irrigation.
3 Make more farmland eg clear forests, reclaim moorland.

6E Solving The Problems of a Slowly Growing Population

Solution 1 – Increase Birth Rates

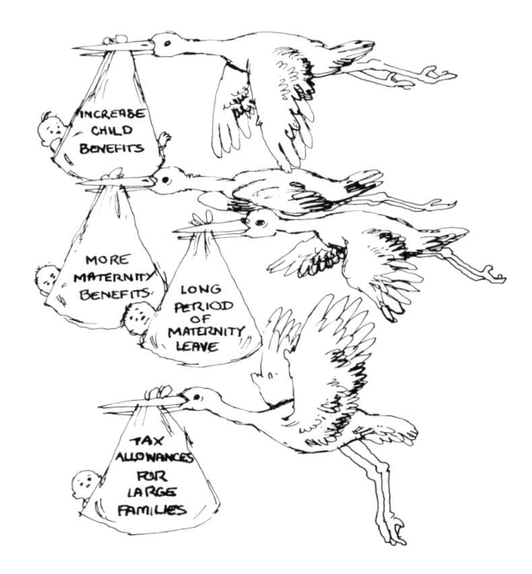

Solution 2 – Increase The Workforce

1 Persuade more women to take jobs eg have crèches in factories.
2 Raise the age of retirement.
3 Increase the number of immigrants.
4 Reduce emigration.

Core Questions

1 Look at 6A.
Which countries have a higher natural increase in population – Developing or Developed Countries?

2 Look at 6B.
Name two problems faced by a rapidly growing country.

Look at 6C.
3 Name one problem faced by a slowly growing country.
4 Which age group makes up the 'active population'?
5 Which age groups make up the 'dependent population'?

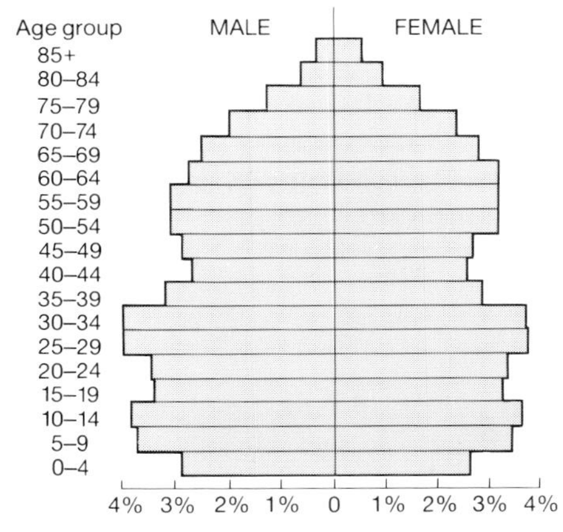

6 Does the population pyramid above show a low or high birth rate? Give one reason for your answer.
7 Does the pyramid show a low or high death rate? Give one reason for your answer.

Look at 6D.
8 Describe two ways in which countries can reduce their birth rate (look at the photograph above).
9 Describe two ways in which countries can develop their wealth.

Look at 6E.
10 Name two ways in which countries can increase their birth rate.
11 Describe how countries can increase their workforce.

Birth control posters in India

Questions

Case Study of India

Look at fig. 6.4.
F1 Is India's population growing rapidly or slowly?

Look at fig. 6.2.
F2 Does India have a lot of old people or very few?
F3 How do you know from the population pyramid that India has a lot of children?

F4 Look at fig. 6.3.
Do you think it is good or bad for India to have so many young people? Give reasons for your answer.

Look at fig. 6.1.
'India's growing population has brought more problems to people in the countryside than in the towns.'

F5 Do you agree with this statement? Give a reason for your answer.

F6 When did India try to develop its wealth – in the 1950s or the 1980s? Give a reason for your answer.

F7 When did India try to reduce its birth rate – in the 1960s or the 1970s? Give a reason for your answer.

F8 Look at fig. 6.1 and 6D.
Do you think India has used the best methods to reduce its birth rate? Give reasons for your answer.

'If the Indian government wants its people to have fewer children, it should advertise on television and in the newspapers.'

Look at fig. 6.5.
F9 Do you agree with the statement above?
Give a reason for your answer.

Questions

Case Study of Germany

Look at fig. 6.6.
G1 In what way does the pyramid show that Germany's birth rate is falling?

Look at figs. 6.6 and 6.2.
G2 Compare the population pyramids of India and Germany.
G3 Explain why the population pyramids of India and Germany are different shapes.

Look at fig. 6.7.

G4 Give one advantage and one disadvantage of Germany having many old people and few young people.

G5 What does the cartoon above tell you about the population problems in Germany?
G6 Parents in Germany can take one year off on maternity leave. Describe one advantage and one disadvantage of this policy.

Raise the retirement age.
Increase the number of immigrants.

G7 Which of the ideas above is the better way of increasing the workforce in Germany? Give reasons for your answer.

Resources

Case Study of India

India's population is growing rapidly because the birth rate is much higher than the death rate. There is not enough land or food for the people in the countryside, so millions move to the cities. In the cities there are not enough houses or jobs, so people live in extreme poverty. The Indian Government has tried many solutions to its population problem:

1950s It developed its coal and iron ore and started many factories.
1960s It grew high yielding crops and made more farmland.
1970s Voluntary and compulsory sterilization.
1980s It set up family planning clinics, advising people on birth control.

Fig. 6.1

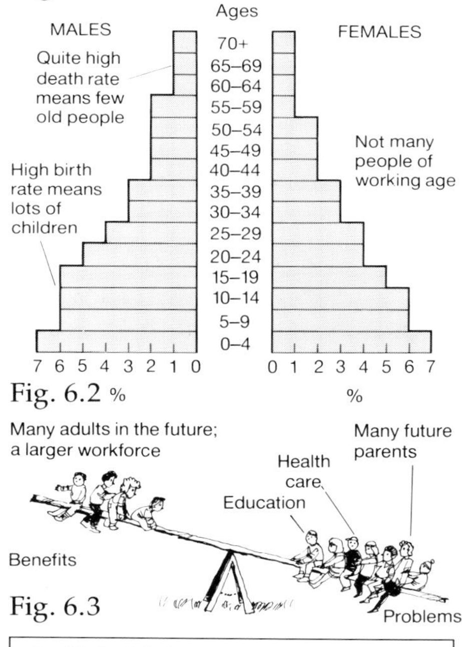

Fig. 6.2

Fig. 6.3

India's birth rate = 32‰
death rate = 14‰

Fig. 6.4

India's literacy rate = 36‰

Fig. 6.5

Case Study of Germany

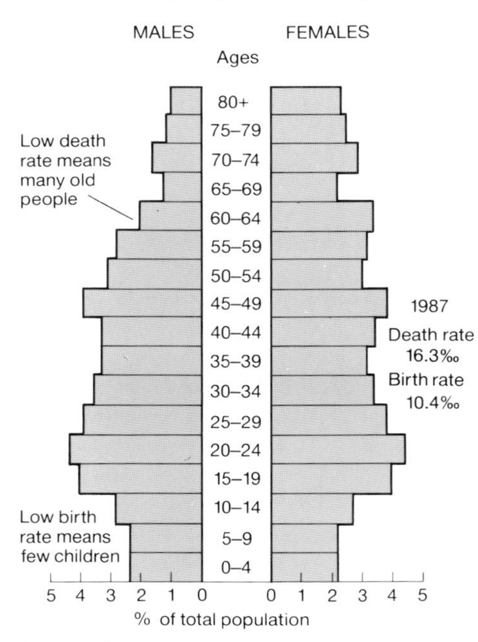

Fig. 6.6

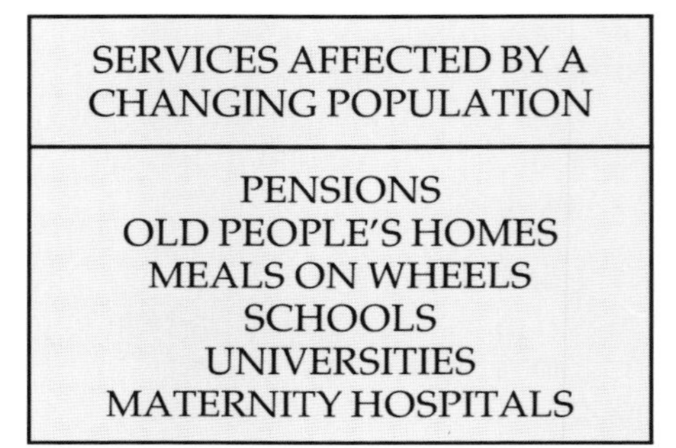

SERVICES AFFECTED BY A CHANGING POPULATION

PENSIONS
OLD PEOPLE'S HOMES
MEALS ON WHEELS
SCHOOLS
UNIVERSITIES
MATERNITY HOSPITALS

Fig. 6.7

Germany has a natural decrease in population because the death rate is slightly higher than the birth rate. This caused a shortage of workers in the 1980s. To solve this problem the Government allowed mothers to take a year off work as maternity leave. People with large families paid less tax. The Government increased the length of conscription (compulsory time in military service) and it allowed in more immigrants and raised the retirement age. Although many of these measures are still in force, the recent unification of East and West Germany means that an increased labour supply is now available.

Fig. 6.8

Case Study of Egypt

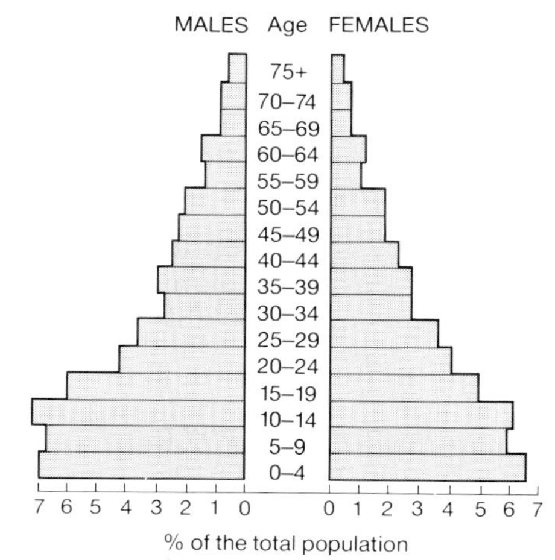

Fig. 6.9

Egypt is 4 times the size of the UK but it has fewer people. It is much more overpopulated than the UK because it has fewer resources. Over 90 per cent of the country is desert and nearly all the people live in a narrow strip along the Nile Valley. Since the 1950s, the Egyptian Government has tried many policies to solve its population problem.

Fig. 6.10

Agricultural Policy – more irrigation schemes and smaller farms.
Mining Policy – develop oil, phosphates and other minerals.
Industry Policy – production has quadrupled eg textiles, food processing.
Tourism Policy – encourage more tourists.
Power Policy – build many power stations.
Family Planning Policy – more family planning clinics and advisers who visit all couples.

Fig. 6.11

Extension Text

6F Overpopulation and Underpopulation

If a country is **overpopulated,** it means there are too many people for the resources available. So the peoples' standard of living is low.

If a country is **underpopulated,** it means there are too few people to develop the resources the country possesses. So the peoples' standard of living will also be low.

If a country has the **optimum population,** it has just enough people to develop its resources fully so that everyone enjoys a high standard of living. The optimum population will not be the same in every country. It depends on the resources of each country.

An overpopulated country can solve its problem by reducing its population and/or improving ways of using its resources eg developing tourism, farming.

An underpopulated country can solve its problem by increasing its population and/or using its resources more efficiently eg using machines instead of workers.

6G Dependency Ratio

The rate at which the population grows affects the population structure. Every country would like a population in which there is a high percentage of active people and a low percentage of dependent people.

The **dependency ratio** shows how 'active' a country's population is.

Country	Dependent Population	Active Population	Dependency Ratio
UK	37%	63%	0.59
USA	41%	59%	0.69
Belgium	42%	58%	0.72
India	48%	52%	0.92
Peru	50%	50%	1.00
Pakistan	50%	50%	1.00

$$\textbf{Dependency ratio} = \frac{\text{\% of dependent population}}{\text{\% of active population}}$$

The lower the dependency ratio, the more 'active' is the population.

Questions

Look at 6F and 6G.

E1 Why does a country have a low standard of living if it is
(*a*) overpopulated, and
(*b*) underpopulated?

E2 Why is it not possible to work out whether a country is overpopulated from its population density?

E3 How do the dependency ratios of Developed and Developing Countries differ?

E4 How is a low dependency ratio of benefit to a country?

Questions

Case Study of Egypt

Look at fig. 6.9.

C1 Describe Egypt's population structure, birth rate and death rate.

C2 Do you think that Egypt's dependency ratio is helpful or harmful to the country's economy?
Give reasons for your answer.

Part of a government debate on Egypt's high birth rate:

Social Services Minister *'Our children will only be useful adults if they are healthy and educated'.*

Employment Minister *'What happens when the children grow up?'*

Agriculture Minister *'There is not enough land for everyone to farm'.*

Finance Minister *'I'm concerned about the number of tax-payers'.*

Defence Minister *'We need a lot of men to defend our country'.*

Read the ministers' viewpoints.

C3 Explain the different points of view that Egyptian ministers have towards a high birth rate.

EGYPT'S POPULATION PROBLEM

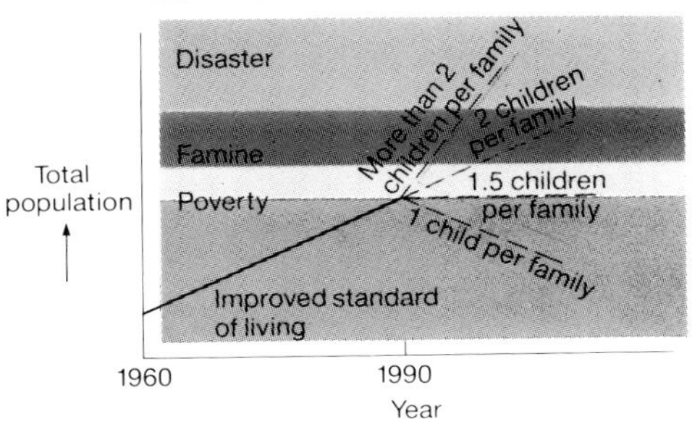

C4 (*a*) What does the graph above show?
(*b*) People disagree about using total population to estimate standard of living. Describe the different points of view people would have.

BIRTH RATES AND CLINICS IN EGYPT

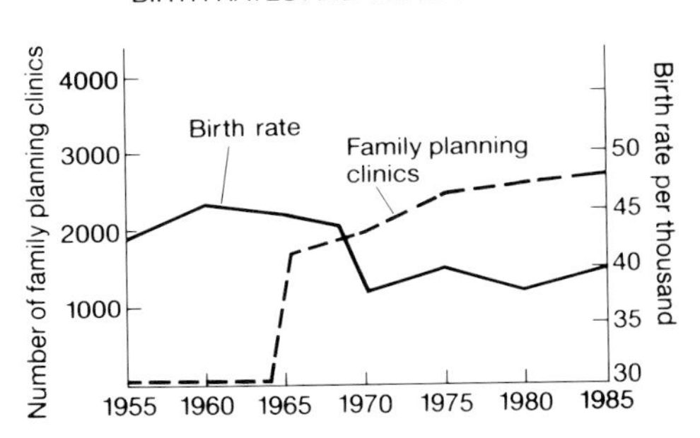

Look at the graph above.

'Our policy of family planning clinics has been a complete failure.' (Egyptian doctor)

C5 The statement above is an exaggeration. Explain how it is exaggerated.

Look at figs. 6.10 and 6.11.

C6 Which of Egypt's five policies (not family planning policy) do you think will do most to reduce overpopulation? Give reasons for your answer.

C7 Choose a suitable approach to family planning that is new to Egypt. Explain your choice and suggest problems that the approach might bring.

UNIT 7 Migration in the Developing World

Core Text

7A Migration

The number of people in a country or region depends upon how many births and deaths there are. It also depends upon how many people move in to the region (**immigrants**) and how many people move away from the region (**emigrants**).

7B Migration from the Countryside To Towns

In Developing Countries, millions of people migrate from the countryside (**rural areas**) to the towns (**urban areas**). Sometimes they move because of the problems in the countryside. These are called **push factors.** Sometimes they move because of the attractions of the towns. These are called **pull factors.**

7C Effects of Migration on The Countryside

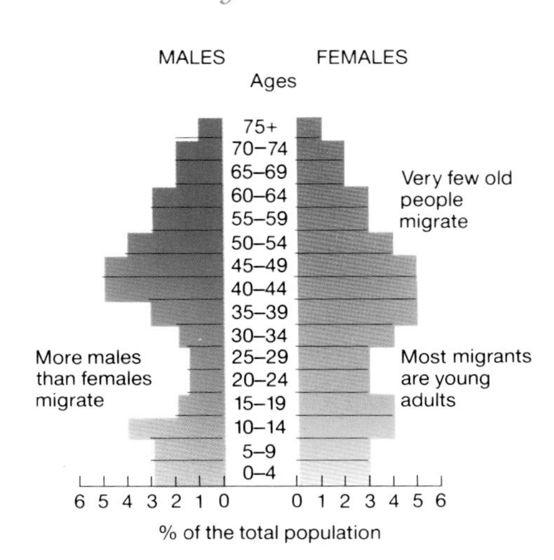

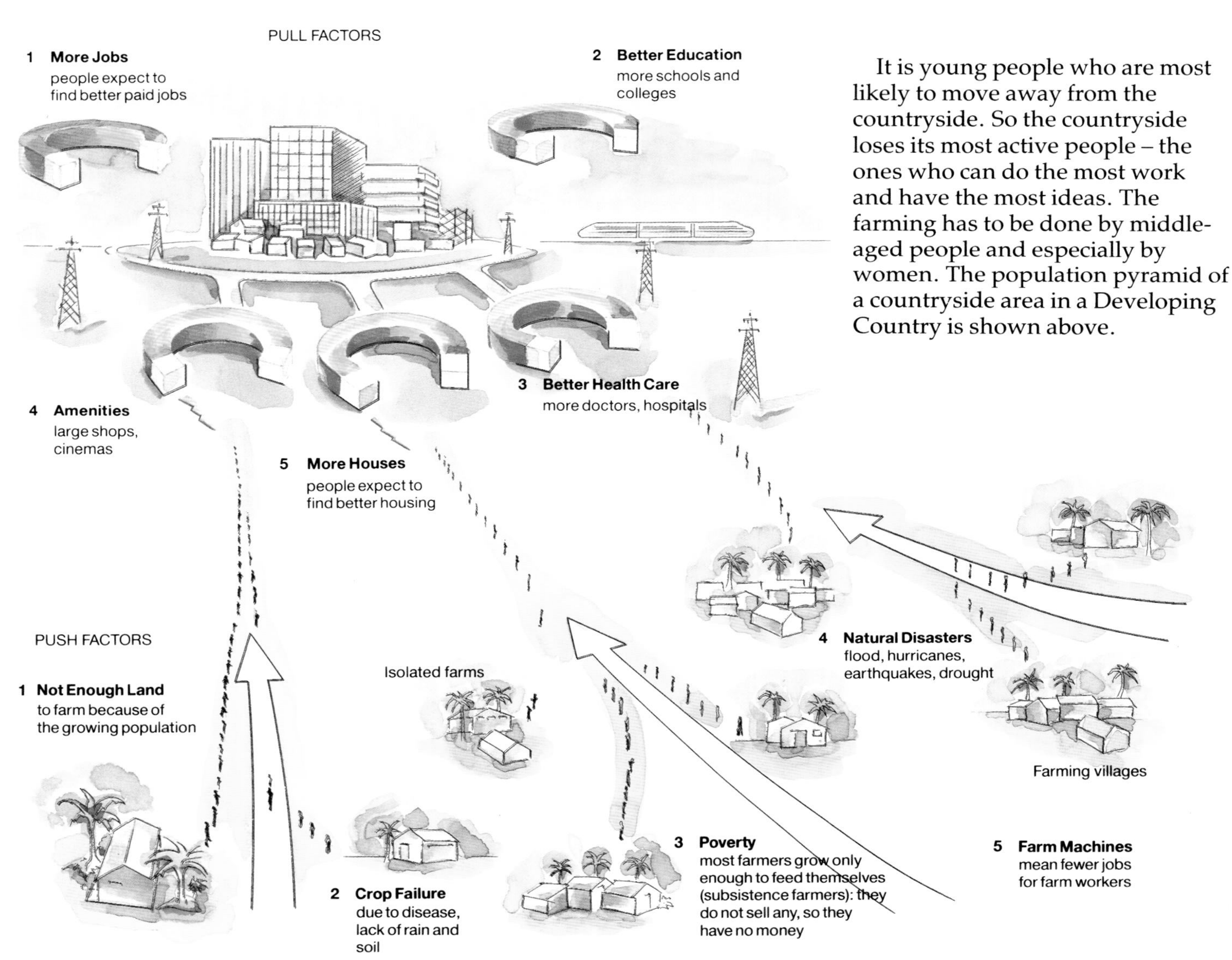

It is young people who are most likely to move away from the countryside. So the countryside loses its most active people – the ones who can do the most work and have the most ideas. The farming has to be done by middle-aged people and especially by women. The population pyramid of a countryside area in a Developing Country is shown above.

7D Effects of Migration on The Towns

Cities in the Developing World are growing very rapidly. This is called **urbanization.** In 1950, the population of Developing World cities was 250 million. By 1980, it had risen to 1000 million and, by the year 2000, should reach 2000 million.

These cities are growing rapidly partly because of the large number of people who are migrating from the countryside. Many cities now have over 1 million people. These are called **millionaire cities.**

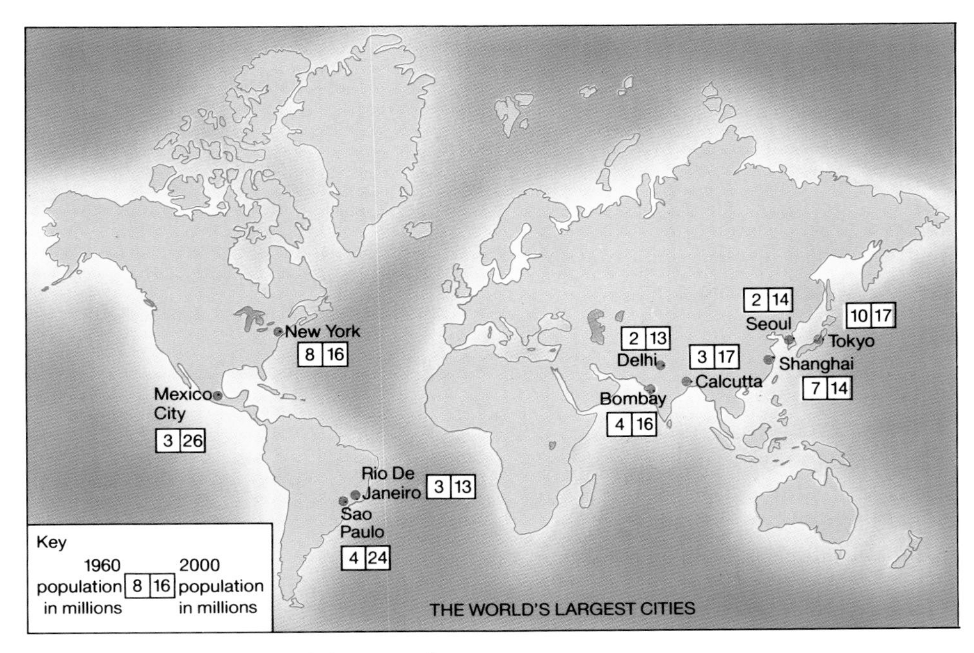

7E Shanty Towns

Cities in Developing Countries cannot cope with so many immigrants. There are not enough jobs or houses. So the people build make-shift houses on waste land. These squatter camps are called **shanty towns,** but they are not separate towns. They are inside the cities. About 40 per cent of the people live in shanty towns. Other names for shanty towns are; *barriadas, favelas, bustees, barrios, ranchos.*

7F Solving the Problems of Migration

1 Build More Low Cost Housing – but even low rents are too much for many people.

2 Improve the Shanties – make them legal, then the people will improve them – provide clean water, sewage disposal and electricity – this is costly.

3 Sites and Services Scheme – provide land with electricity, water and other services laid on – squatters build their own homes with loans from the council.

4 Build Industrial Estates – provide land with services laid on – people can then start businesses eg tradesmen.

5 Improve Public Transport – so people can find work in other parts of the city.

6 Improve Life in the Countryside – more health centres, schools, better farming – then fewer people will want to leave.

Core Questions

Look at 7A.

1 What is the difference between immigrants and emigrants?

Look at 7B.

2 What is a rural area?

3 What is an urban area?

4 Describe three 'push factors' which cause people to move from the countryside.

5 Describe three 'pull factors' which attract people to the cities.

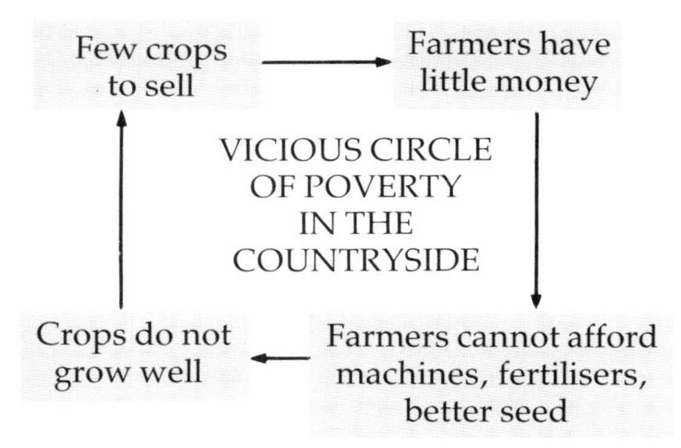

6 What does the diagram above show?

Look at 7C.

7 Which age groups are most likely to emigrate from the countryside?

8 Do more females or males emigrate from the countryside?

Look at 7E.

9 What are 'shanty towns'?

10 Describe what it is like in a shanty town.

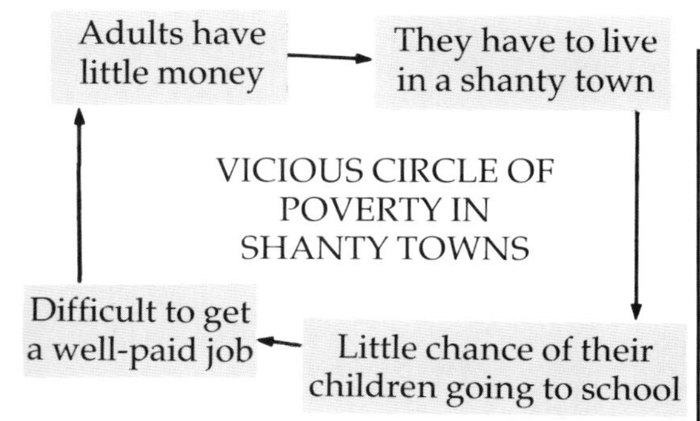

11 What does the diagram above show?

Questions

Look at 7D.

E1 Compare the growth of cities in the Developing and Developed world between 1950 and 2000.

E2 What is the cartoon above trying to show?

Look at 7E.

E3 People have different views towards shanty towns. Identify and describe these different points of view.

Look at 7F.

E4 Which method (1-6) will do most to solve the problems of Developing World cities? Give reasons for your answer.

ARTICLE A

A Shanty Town in Sao Paulo

The people live in shacks of wood or corrugated iron. Out houses serve as sewers. Rubbish, stench, disease and decay are everywhere. Life expectancy is short here. Many babies never even live to see childhood. Over half of Sao Paulo's population live in conditions like this.

ARTICLE B

Tourist Information on Sao Paulo

Don't miss Sao Paulo. It is fast, noisy, impressive and exciting. The houses are remarkable for their beauty; the beaches are immaculate and the shops luxurious. This is Brazil's most prosperous city.

E5 In what ways is article B above exaggerated, and why is it exaggerated?

Core Investigation

The next two pages give information about migration in Peru.

Write a report on migration in Peru.

The aim of your investigation is to find out:

(*a*) why people in Peru emigrate from the countryside.

(*b*) why people in Peru migrate to the city of Lima.

(*c*) what problems this migration brings to Lima.

Your answer should be in four parts.

1 Aim 3 Analysis
2 Methods 4 Conclusion

Use other books and atlases to get more information.

Look at page 16 to find out how to write up your report.

Core Discussion

Your group represents a shanty town committee in Lima.

You have been given £2 million by the World Bank to spend on your shanty town. Discuss how you will use the money. Present your plans to the class giving reasons for the choices you made.

The spending choices you have are given at the bottom of page 38.

Case Study of Migration in Peru

Problems in the Peru Countryside

5% of Peru is fertile farmland
40% has poor soils covered with rainforest
30% has thin soils and steep slopes
15% is too cold to grow crops
10% is too dry to grow crops.

A TYPICAL MOUNTAIN VILLAGE IN PERU

Housing – made of mud-brick called **adobe:** no electricity, no water: heating from fires, using animal dung as fuel

Sewage – no toilets – sewage makes its way into rivers

Water supply – from the nearest river, which is often polluted with sewage – this spreads disease

Medicine – no doctor or clinic

School – a primary school, but children do not go if there is work to do on the farm – a secondary school is in the nearest town, but it has to be paid for

Natural Disasters – earthquakes occur, causing landslides and avalanches: droughts are common.

It is the custom of the farmer, when he dies, to divide his land between his sons. He may have several sons.

Until recently, there were many large estates (**latifundio**) owned by absentee landlords who spent their profits elsewhere. The peasants worked the land and were paid a very low wage. They were given tiny plots of land on the estate to farm, but they had to pay the landlord rent in the form of crops. If they grazed their animals on the estate they were paid no wage at all.

Food Production Per Person in Peru	
1965	100
1970	105
1975	97
1980	82
1985	77

1965 food production is taken as a base

'Poverty remains a problem (in the Peru countryside), causing serious diseases, partly from lack of food, partly because the food includes very little animal or vegetable protein. (Morris, 1979)

Most of the trees on the hillsides have been cut down. With no roots to stabilise the soil, it is eroded by the rain, leaving deep gullies in the hillside.

A mountain village in Peru

The Attractions of Lima, The Capital City

	Countryside	Lima
People with safe water	25%	60%
People with sewage disposal	17%	51%
Homes with electricity	3%	55%
Infant mortality	58‰	73‰
Weekly income	300 soles	1500 soles (in shanty town)

Lima is easily the biggest city in Peru, with shopping centres, a university, schools, hospitals and recreational areas. Seventy per cent of all Peru's manufacturing industry is here and most of the wealthy people live here in rich suburbs.

Most immigrants come from the mountains. They are usually young and better educated than most of the people in the countryside. Many migrate to smaller towns before coming to Lima.

The Growth of Lima	
1940	500 000
1950	900 000
1960	2 000 000
1970	3 200 000
1980	5 500 000
1990	8 000 000

One third of all the people in Peru live in Lima.

One third of all the people in Lima live in shanty towns, called barriadas.

A TYPICAL SHANTY TOWN IN LIMA

Home – small shack made of scrap wood, matting or tin; no electricity; no water, no sewage; no refuse collection.

Roads – dust tracks.

Food – too poor to buy much food; most people are malnourished.

Health – no doctors or clinics; water is polluted with sewage; diseases common eg typhoid, dysentery.

Shanty town in Lima

Lima is in a desert. It depends for its water on the River Rimac. This was not enough, even in the 1960s. Since then the population of Lima has quadrupled. People in shanty towns buy their water at 40 times the cost of piped water.

Sixty per cent of the people of Lima work in service industries and half of these do not have regular work. Only 17 per cent work in manufacturing, a sign that the manufacturing industry cannot cope with the growing population.

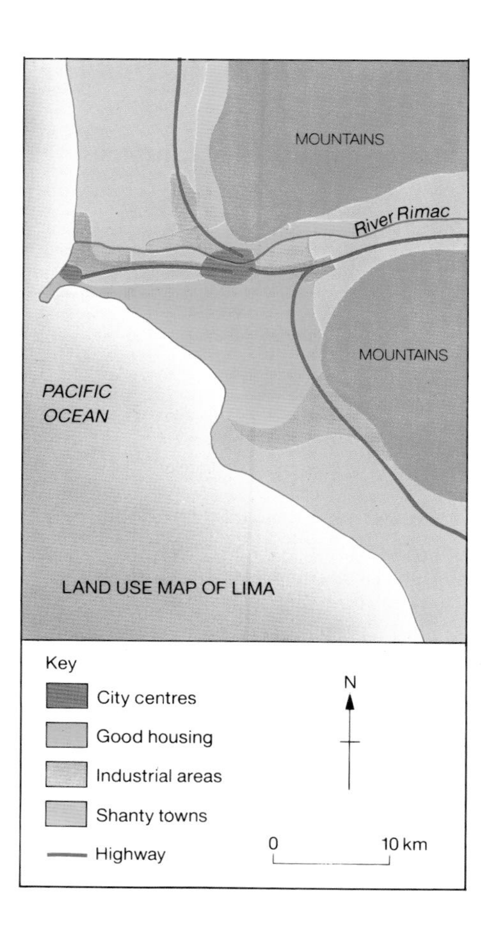

How would you invest the £2 million given for shanty town improvements?

	Cost
Primary School	£200 000 each
Secondary School	£400 000 each
Hospital	£1 500 000 each
Doctor	£200 000
Clean water for everyone	£250 000
Clean tap in each street	£50 000
Electricity in every house	£400 000
Connect every home to a sewer	£100 000
Bulldoze shanty town	£250 000
Build low cost housing	£2 million
Build railway to city centre	£500 000
Make new bus service to city centre	£50 000
Build industrial estate	£1 million
Surface all roads	£500 000

UNIT 8 International Migration

Core Text

8A International Migration

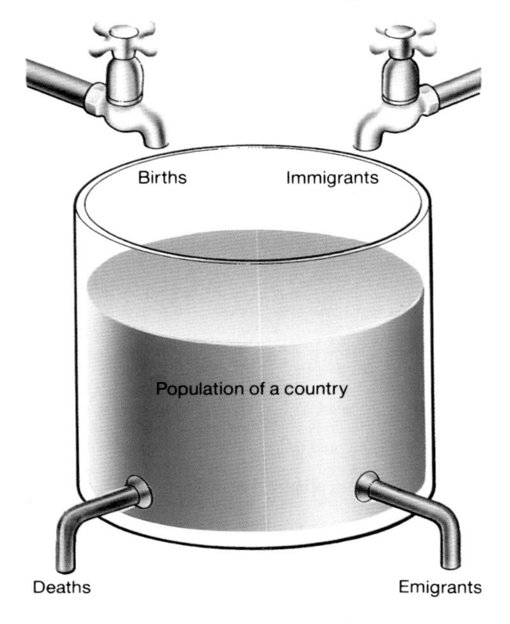

A country grows in population through **births** and **immigration**. A country loses people through **deaths** and **emigration**.

When people move from one country to another it is called **international migration**. They may move because of **push factors** (things they do not like in their own country) or **pull factors** (things that attract them to another country).

Sometimes people emigrate to another country because they have to (called **forced migration**). Often people emigrate because they want to (called **voluntary migration**).

8B Short- and Long-Term Migration

Short-term migration is when people move for a few months or a few years and then return home. They usually do this to earn money. They are sometimes called **guest workers.**

Long-term migration is when people move permanently. They do not return home.

8C Reasons for Voluntary Migration

8D Effects of Migration

On the losing country:

1 Lose a lot of 20-35 year olds and their children.
2 Lose go-ahead people.
3 Need to provide fewer jobs and services and less food.
4 Lose more males, causing family break-ups.

On the gaining country:

1 A bigger labour force.
2 Need to provide more services for the extra people.
3 Often leads to shanty towns in cities.
4 Different races and religions live side by side.

8E Forced Migration

People sometimes migrate because they are treated badly. This might be due to their **religion**, **race** or **politics**. They may also migrate if there is a **war** or a **natural disaster**, eg famine or earthquakes.

People who have been forced to move away from their country are called **refugees.** There are about 10 million refugees in the world and half of them are children. Some find a better life in their new country, but many live in **refugee camps** in extreme poverty.

A refugee camp

Core Questions

Look at 8A.
1 What two things cause the population of a country to grow?
2 What two things cause a country to lose people?

Look at 8B.
3 What are guest workers?

Look at 8C.
4 Give two reasons why people move away from their country.
5 Give three reasons which attract people to another country.

Look at 8D.
6 Describe two effects on a country which 'loses' a lot of emigrants.
7 Describe two effects on a country which 'gains' a lot of immigrants.

Questions

Case Study of Jamaica

Look at fig. 8.2.
F1 Which of the Jamaican migrations (A,B, or C) is an example of forced migration? Give a reason for your answer.

Look at fig. 8.1.
'Conditions in Jamaica make many people want to leave'
F2 Do you agree with the statement above? Give reasons for your answer.

Look at figs. 8.4. and 8.5.
F3 Why do you think Jamaicans have been attracted to the UK?

Look at fig. 8.3.
F4 How has the number of West Indian immigrants to the UK changed since 1951?

'We shall not let young Jamaicans emigrate to the UK. We need them here' (Jamaican official)
F5 Do you agree with this point of view? Give reasons for your answer.

Look at fig. 8.5 and 8.7.
F6 How has the immigration of Jamaicans helped Britain?

Look at fig. 8.9.
F7 Describe the problems that Jamaican immigrants to Britain have to face.

Questions

Case Study of Turkey

Look at fig. 8.11.
G1 Which of the migrations to and from Turkey have been (*a*) forced (*b*) long-term voluntary (*c*) short-term voluntary?
G2 Were the people emigrating from Bulgaria in 1989 refugees? Give a reason for your answer.

Look at fig. 8.12.
G3 Describe how immigration and emigration in Turkey have changed in the 1980s.

Look at figs. 8.10 and 8.13.
G4 Some people say that people emigrated from Turkey to Germany because of 'push factors'. Others say it was because of 'pull factors'. Give reasons to support both points of view.

Look at fig. 8.16.
G5 Do you think that Turkey is pleased that so many of its people emigrate as guest workers? Give reasons for your answer.

Look at fig. 8.15.
G6 Describe how Germany's attitude towards immigrants has changed since the 1960s.

Look at the cartoon to the right.
G7 What is the cartoon trying to say about the problems of Turkish immigrants in Germany?

Look at fig. 8.17.
G8 The emigration of Turks to Germany causes problems to both countries. Which of the two schemes will do more to reduce this problem? Give reasons for your answer.

Resources

Case Study of Jamaica

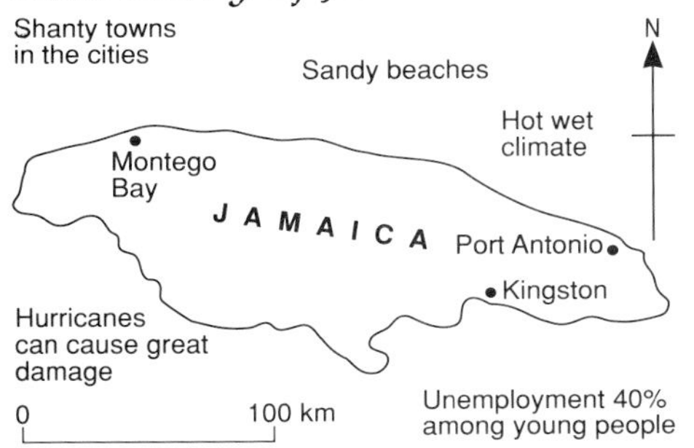

Fig. 8.1

A. 17th Century: Immigration to Jamaica
Many Africans were sent to Jamaica as slaves to work on the sugar plantations.

B. 19th Century: Immigration to Jamaica
Slavery was abolished and many Indians came to Jamaica to find work on the plantations.

C. 20th Century: Emigration from Jamaica
Many Jamaicans left to find a better way of life abroad.

Fig. 8.2

EMIGRANTS FROM THE WEST INDIES TO THE UK

Year	*Number*
1951	1 000
1956	26 000
1961	65 000
1966	14 000
1971	5 000
1976	4 000
1981	4 000
1986	5 000

Fig. 8.3

ADVERTISEMENT IN JAMAICA IN THE 1960s

LONDON TRANSPORT REQUIRES BUS CONDUCTORS IMMEDIATELY
Successful applicants could become trained drivers. Good housing is provided by the boroughs. Schools are of the highest quality.
COME AND MAKE A NEW LIFE FOR YOURSELF

Fig. 8.5

'Without black workers the hospitals, buses and trains in London would probably come to a halt.' Fig. 8.7

Jamaican immigrants arriving in the UK

Fig. 8.8

Conditions in a Jamaican city

Fig. 8.6

	UK (1980)	Jamaica (1980)
Income per person	$9110	$1180
Calories per day	3343	2624
Children at secondary school	82%	57%
Population per doctor	650	2800
People in agriculture	2%	21%

Fig. 8.4

Inquiry finds that black people in Britain are unfairly treated

Black people in parts of London given poorer housing than whites

Unemployment amongst people of Jamaican origin twice that of the white population in Britain

Fig. 8.9

Case Study of Turkey

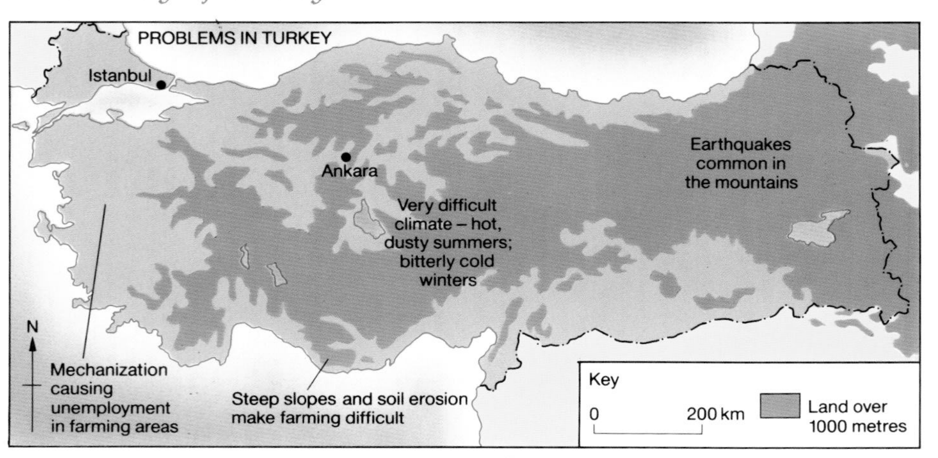

Fig. 8.10

1920s 1 million Greeks decide to leave Turkey. 500 000 Turks choose to return home from Greece.
1940s 250 000 Turks told to leave Bulgaria and return to Turkey.
1960s 250 000 Turks leave to work in West Germany as guest workers.
1980s 300 000 Turks, persecuted because of their religion in Bulgaria, return to Turkey.
Turks leave to work in the Middle East as guest workers.

Fig. 8.11

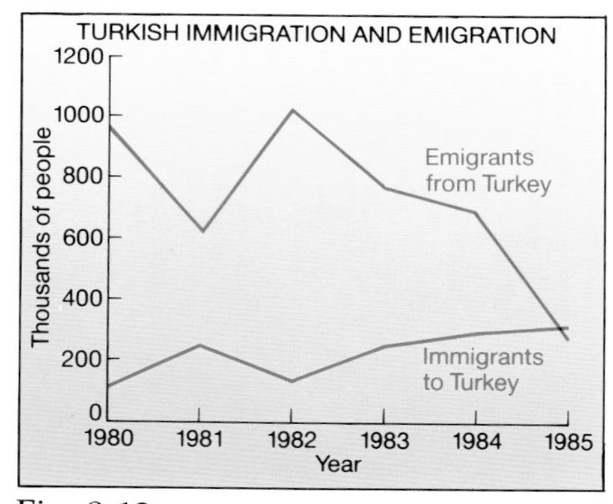

Fig. 8.12

	Turkey	Germany
Income per person	$1500	$12000
Population per doctor	2850	450
People who can read and write	68	99
TV sets per thousand people	12	250

Fig. 8.13

Fig. 8.14 *Rural Turkey*

In the 1960s, West Germany wanted immigrants because they were willing to do the low paid, dirty, unskilled jobs that German workers would not do. In the 1980s unemployment in West Germany was much higher. Germany offered Turkish guest workers £1000 each to return home.

Fig. 8.15

Unemployment 20% in Turkey

Difficult for Turkey to develop with so few 20-40 year olds

Guest workers send home £1 million each year to their families in Turkey

Turkey restricts the number of immigrants from Bulgaria.

Turkey wants more people to live in the east, to defend its borders

Population of Turkey growing at 2.5% each year

Fig. 8.16

B DEVELOPMENT OF RURAL TURKEY

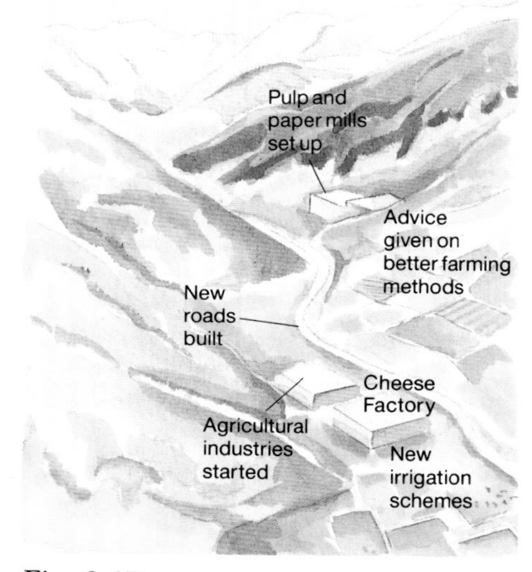

Fig. 8.17

Case Study of Hong Kong

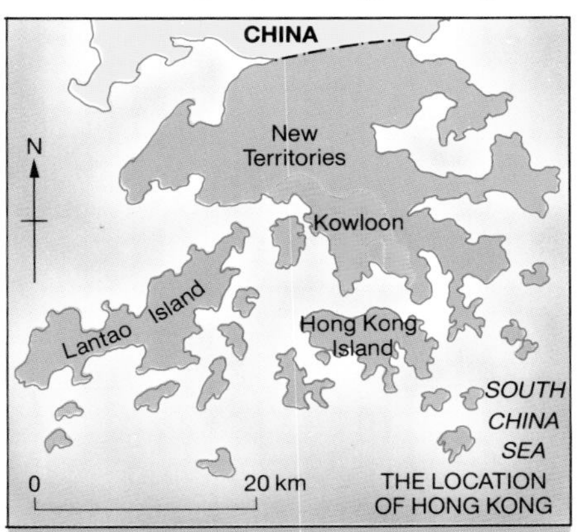

Population density of Scotland: 65 per sq km.

Population density of Hong Kong: 2459 per sq km.

When immigration to Hong Kong began, empty houses were divided up so that up to 100 people lived in a single house.

500 000 people live in shanty towns or on rooftops or in boats.

Hong Kong is trying to discourage the emigration of its computer programmers.

Hong Kong is short of water. It has had to build an expensive desalination plant.

There are a quarter of a million job vacancies in Hong Kong. It has been suggested that workers be brought in from China.

Many thousands are expected to emigrate from Hong Kong before it is taken over by China in 1997.

In 1989 cholera broke out in one of the Vietnamese detention camps.

Fig. 8.18

	Vietnam (1980)	Hong Kong (1980)	
GDP/head	$98	$5000	1975: Communists overrun Vietnam. Millions flee to Hong Kong.
% in agriculture	71	3	
% natural increase	3	2.3	1983: 4 typhoons in Vietnam leave 650 dead and ruin harvests
Population per doctor	4190	1220	
Calories eaten as a % of those needed	90	128	1985: typhoon kills 670 and ruins farmland in Vietnam
Adult literacy	87	90	1987/88: food shortages in Vietnam
Unemployment	30%	2%	

Fig. 8.20

1940s/50s 700 000 flee Communist China into Hong Kong.

1950s 50 000 squatters rehoused after fire in shanty town.

1970s/80s 120 000 Vietnamese flee Communist Vietnam. Many resettled in western countries eg USA, Canada, Australia.

1979 Some shanty towns close down. 30 000 squatters move to permanent housing.

1987 60 000 professional people emigrate from Hong Kong.

1989 51 Vietnamese in Hong Kong repatriated to Vietnam. 100 Vietnamese volunteer to return to Vietnam.

Fig. 8.21

SOLUTIONS TO THE PROBLEMS OF VIETNAMESE EMIGRATION TO HONG KONG

A **Forcible repatriation to Vietnam**
Vietnamese in Hong Kong told they must return. Vietnam given money for every person returned to their country.

B **Improve conditions in Vietnam.**
Western countries eg USA, UK and France could send aid to Vietnam to improve people's standard of living.

C **Other countries accept Vietnamese Immigrants**

Fig. 8.22

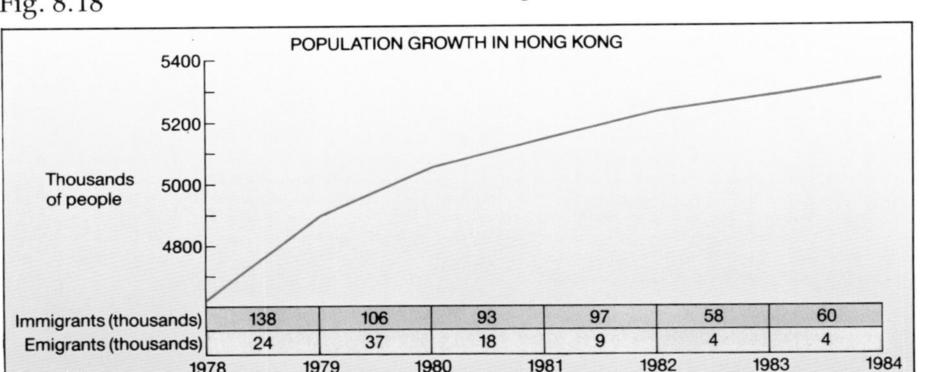

Fig. 8.19

Vietnamese repatriation from Hong Kong

Extension Text

8F Voluntary Migrations

There are many different types of migration. They may be **forced** or **voluntary.** They may be **short-term** or **long-term**, and they may be **short distance** or **long distance.**

Regional	International
Daily	
commuters	
Seasonal	
holidaymakers	holidaymakers
migrant workers	migrant workers
transhumance farmers	nomadic herders
Short-term	
rural → urban	guest workers
urban → rural	
urban → urban	
Long-term	
rural → urban	to nearby countries
urban → rural	across continents
urban → urban	

Voluntary migrations take place for different reasons. They may be for **personal/social reasons** eg a desire to live in the countryside, to be with relations. They may be for **economic reasons** eg to find a job, to explore new territories. Nomads and transhumance farmers migrate in order to make the best use of the available land. Some migrations are for **political reasons** eg a government may encourage people to move to a New Town. They may be for **religious reasons** eg the Jewish migration to Israel after World War 2.

8G Forced Migrations

Forced migrations take place for different reasons. They may be for **political reasons** eg Vietnamese fleeing from Communist Vietnam, people escaping from wars. They may be for **racial reasons** eg segregation of black people into townships in South Africa. They may be for **economic reasons** eg Highland Clearances. They may be for **environmental reasons** eg Ethiopians forced to move because of famine. And they may be for **religious reasons** eg the movement of Jews during World War 2.

Regional	International
urban redevelopment	Slave Trade
job transfers	deportation of criminals
apartheid 'townships'	natural disasters
natural disasters	

Questions

Look at 8F.

E1 Give two examples of voluntary seasonal migration.

E2 Give two reasons why people make voluntary migrations.

Look at 8G.

E3 Give two examples of forced regional migration.

E4 Give three reasons why forced migrations take place.

E5 Draw the table below. Complete the table for each of the migrations listed below.

1 Irish families emigrating to the USA after the potato famine in the 1840s.
2 African Muslims making a pilgrimage to Mecca in Saudi Arabia.
3 Greek women joining their husbands who have emigrated to Australia.
4 The transfer of Civil Service jobs from London to Scotland.
5 Race horse jockeys going to a different racecourse every day.
6 Asians expelled from Uganda.
7 British engineers with two year contracts in Saudi Arabia.

Questions

Case Study of Hong Kong

Look at fig. 8.19.

C1 Describe how immigration and emigration have affected the population growth in Hong Kong.

C2 Draw a table similar to that in question E5. Complete the table for the main migration movements to and from Hong Kong, using the information in fig. 8.21.

Look at fig. 8.20.

C3 People disagree as to whether Vietnamese immigrants to Hong Kong have been political refugees.
Give reasons to support the different points of view.

Look at fig. 8.18.

C4 Describe the advantages and disadvantages of large-scale immigration to Hong Kong.

Look at fig. 8.22.

C5 Which alternative is the best solution to the problem of Vietnamese emigration to Hong Kong? Give reasons for your answer and predict the possible consequences.

'To lose so many ambitious 20–35 year olds must do immense harm to Vietnam.'

Look at fig. 8.20.

C6 Do you agree with the statement above? Give reasons for your answer.

Number of example	Voluntary/ forced	Regional/ international	Short-term/ long-term	Reason

UNIT 9 Clearing the Tropical Rainforest

Core Text

9A Tropical Rainforests

All countries try to improve the standard of living of their people. This is called **development**. For countries beside the Equator, one way to develop is to clear their large areas of tropical rainforest. But clearing the rainforests brings problems as well as benefits.

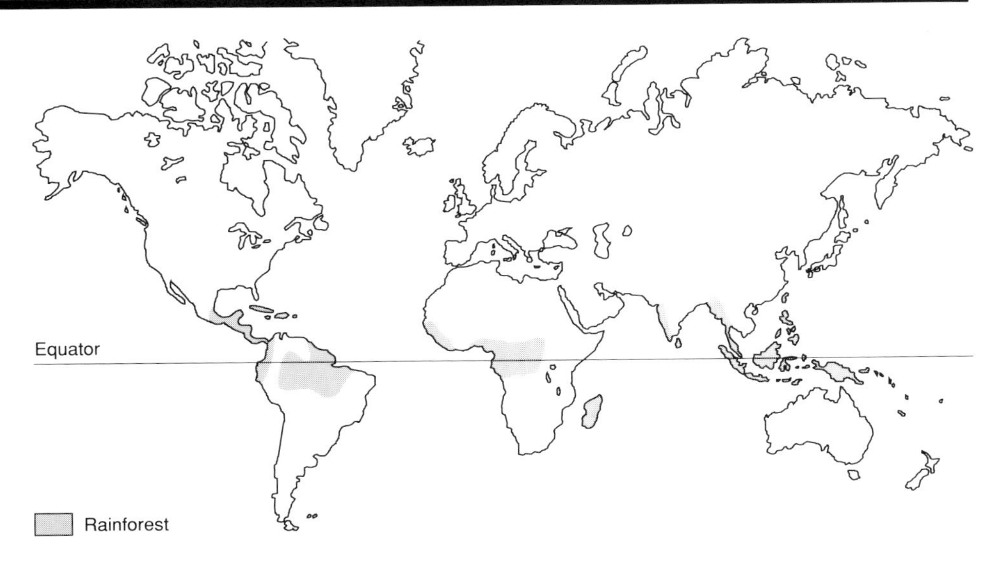

9B Why Rainforests Are Cleared

1 To make farmland for growing crops or for cattle ranching
2 For mining
3 To build roads
4 To build reservoirs
5 To sell the trees as hardwoods or to use them for fuel.

9C Where Rainforests are Being Cleared

AFRICA
REST OF CENTRAL AND SOUTH AMERICA
REST OF SOUTH-EAST ASIA
THAILAND
MALAYSIA
MEXICO
COLOMBIA
BURMA
INDONESIA
BRAZIL

0 25 50 75 100

Thousands of square kilometres of rain forest cut down in 1989

9D Effects of Clearing Rainforests

GLOBAL EFFECTS
Greenhouse effect

EFFECTS ON THE WILDLIFE
Forests are home to millions of birds and animals
If the trees disappear, so does the wildlife

EFFECTS ON THE LOCAL PEOPLE
Millions of people live in the tropical rain forests
Many people hunt animals, collect fruits and make small clearings to farm
With no trees, people's way of life is destroyed

EFFECTS ON THE SOIL
When trees are cleared, there are no roots to hold the soil together
When trees are cleared there are no leaves and roots to make the soil fertile
On slopes the soil is washed away by rain
River is filled with soil and floods often

Core Questions

Look at 9B.
1 Give three reasons why tropical rainforests are being cleared.

Look at 9C.
2 Which three countries cut down the most tropical rainforest in 1989?

Look at 9D.
3 How many people live in the tropical rainforests?
4 What happens to these people when the tropical rainforests are cleared?
5 Why are so many animals and plants of the rainforest becoming extinct (dying out)?

Look at the headline above.
6 Why has this law been made in some rainforest countries?
7 Areas of cleared rainforest are not very fertile. Why is the soil less fertile when the rainforest is cleared?

Questions

Case Study of the Indonesian Rainforest

Look at fig. 9.6.
F1 Did Indonesia clear more or less of its rainforest in 1990 than in 1980?

Look at fig. 9.2.
F2 Which two islands of Indonesia are most crowded?
F3 Which two islands have the most rainforest?

A cleared rainforest

Look at fig. 9.4.
F4 Give two advantages of clearing the Indonesian rainforest.

Look at fig. 9.5.
F5 How does clearing the rainforest affect wildlife?

An Indonesian tribe in the rainforest

'Clearing the forest benefits everyone in Indonesia'

F6 Do you agree with the statement above? Give a reason for your answer.

Look at 9D.

'Clearing the rainforest provides more rich farmland for our people' (Government official)

F7 The statement above is exaggerated. Explain how.

Look at fig. 9.3.
F8 Which solution (A, B, or C) is the best way of protecting the people and wildlife of the Indonesian rainforest? Give reasons for your answer.

Look at fig. 9.8.
F8 Which solution (X or Y) will do more to reduce the amount of rainforest being cleared? Give reasons for your answer.

Questions

Case Study of the Amazon Rainforest in Brazil

Look at fig. 9.11.
G1 Describe the advantages of clearing the Amazon rainforest.

Look at fig. 9.12.
G2 Compare the amount of rainforest cleared in the different regions of Amazonia since 1974.

Look at fig. 9.10.
G3 Give one advantage and one disadvantage to the Indian people of clearing the Amazon rainforest.

Look at figs. 9.11. and 9.14.
G4 There are arguments for and against clearing the Amazon rainforest. How would the arguments of a conservationist and a government minister differ?

G5 The woman's statement in the diagram above is exaggerated. Explain how it is exaggerated.

Rainforest burger

G6 What is the diagram of the 'rainforest burger' trying to show?

Look at fig. 9.13.
G7 Which of the solutions (A, B or C) will most slow down the clearing of the Amazon rainforest? Give reasons for your answer.

Resources

Case Study of the Indonesian Rainforest

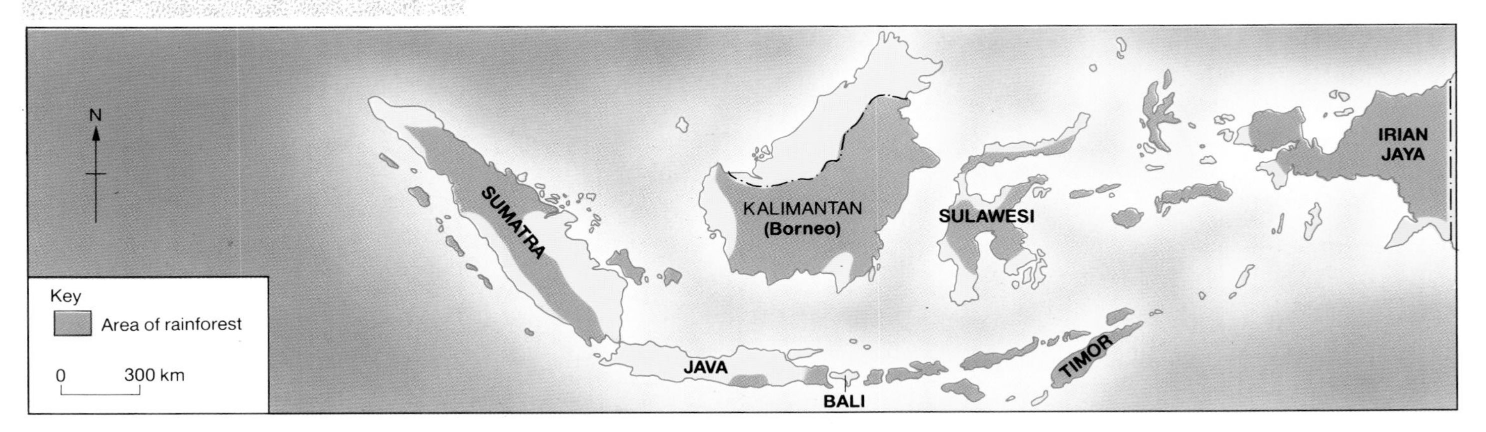

Fig. 9.1

Islands of Indonesia	Population Density (people per sq km) 1980	Amount of Forest (million ha)
Sumatra	71	25
Kalimantan	15	37
Sulawesi	63	11
Irian Jaya	3	29
Java	758	<1
Bali	486	<1

Fig. 9.2

A Set up game reserves in the forest to protect rare animals.

B All timber companies must replant the areas they clear.

C Forests where tribes live should not be cleared.

Fig. 9.3

Indonesia is the fifth most populated country in the world. Most of the country is covered with tropical rainforest. The government wants to clear the forest to make more farmland. It can then export the timber, and people can move from the overcrowded parts of the country.

Fig. 9.4

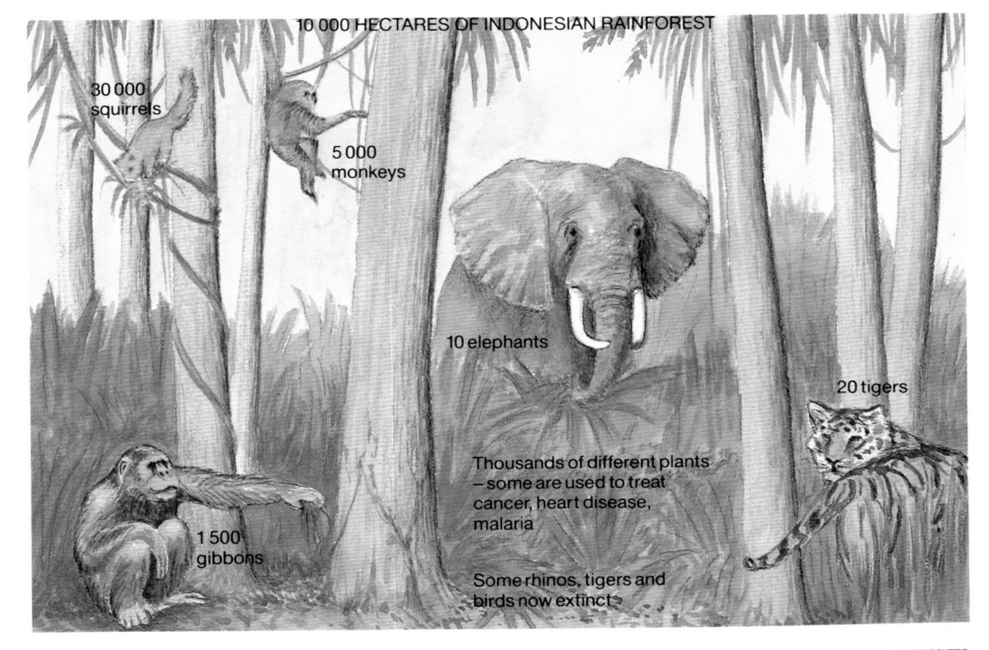

Fig. 9.5

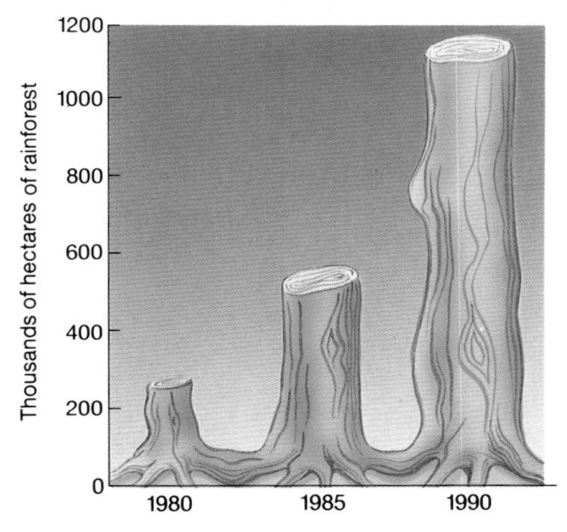

Fig. 9.6

A village in the Sumatran rainforest

Fig. 9.7

X Other countries refuse to buy wood from Indonesia.

Y Other countries refuse to lend Indonesia money to clear its forest.

Fig. 9.8

Case Study of the Amazon Rainforest in Brazil

Brazil is the sixth most populated country in the world. Forty per cent of the country is rainforest, drained by the River Amazon and its tributaries. The Brazilian Government is rapidly developing its rainforest.

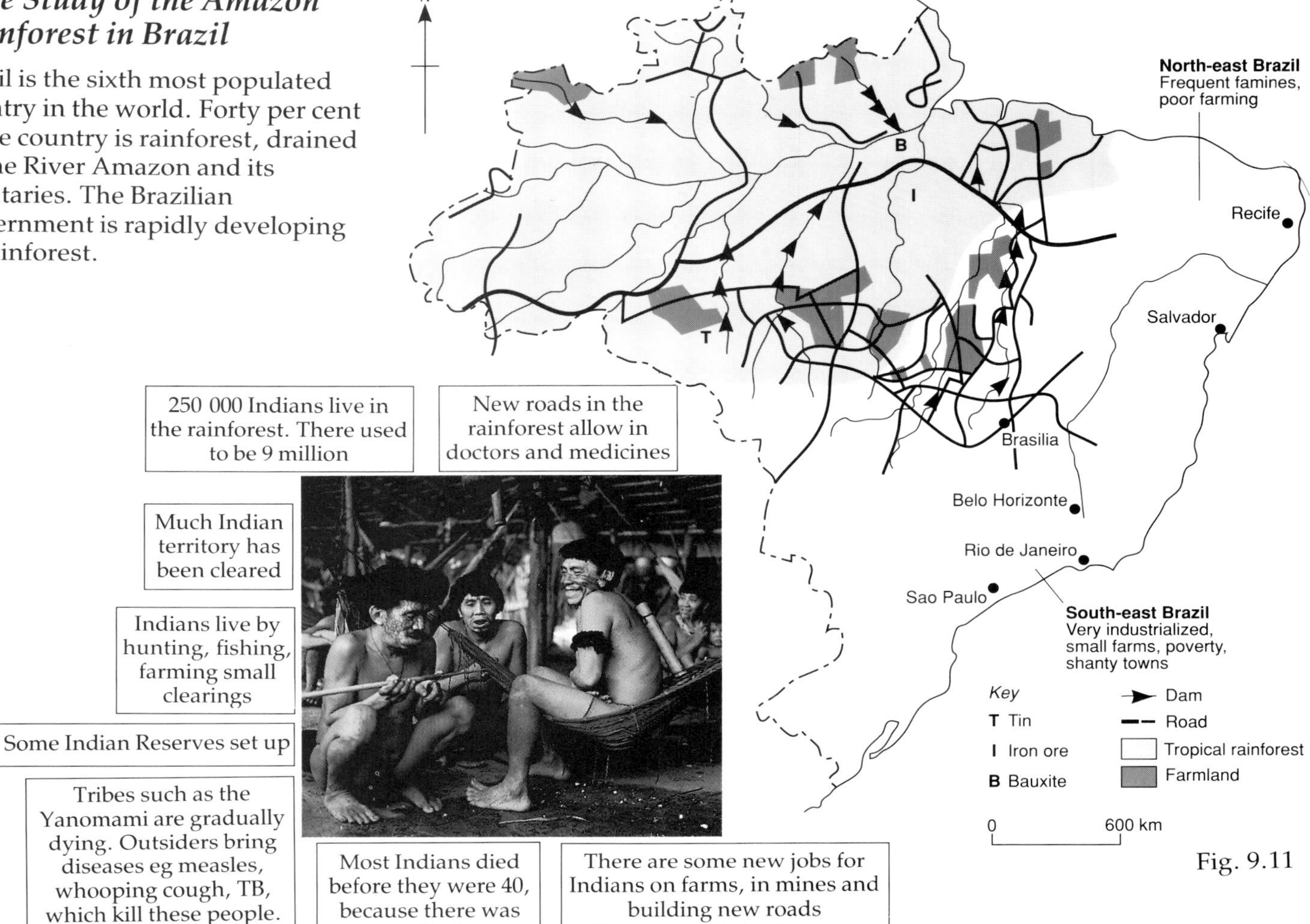

Fig. 9.10

Fig. 9.11

Amazonia	Area cleared of Rainforest 1974	1981	1988
East	2%	4%	7%
Central	1%	5%	10%
West	1%	3%	14%

Fig. 9.12

A Hamburger Company refuses to buy meat from the Amazon rainforest.

B World Bank stops funding new roads in the Amazon rainforest.

C Landowners in Amazonia only allowed to clear forest from half of their land.

Fig. 9.13

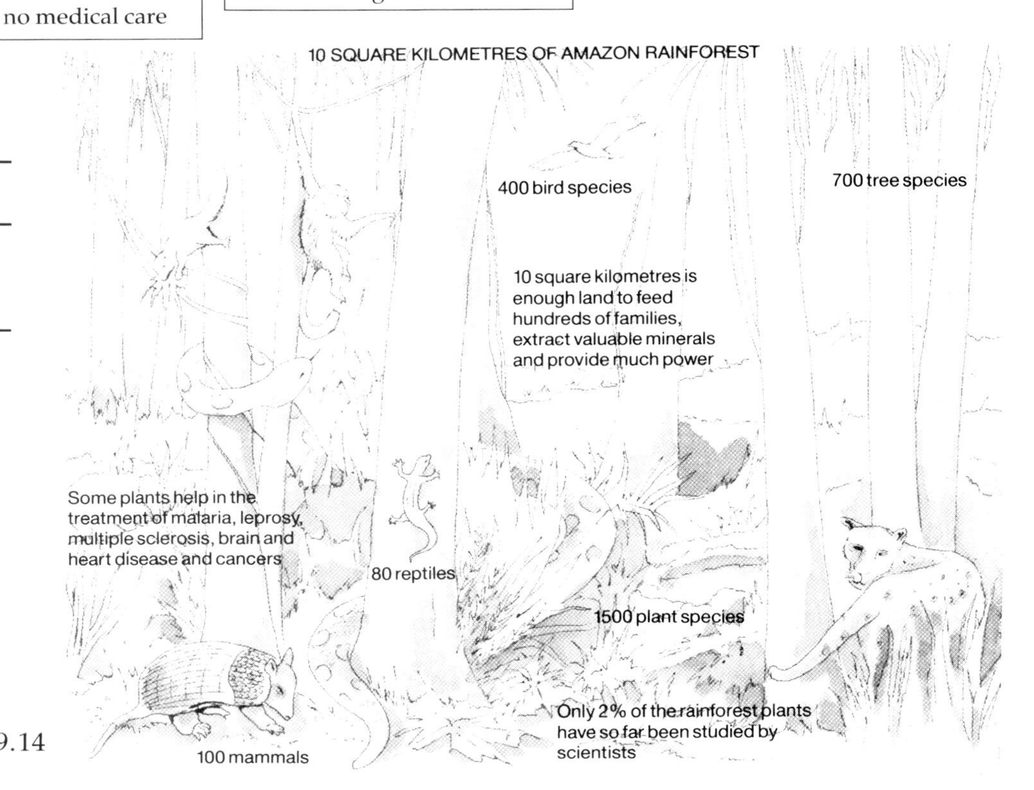

Fig. 9.14

Case Study of the Central American Rainforest

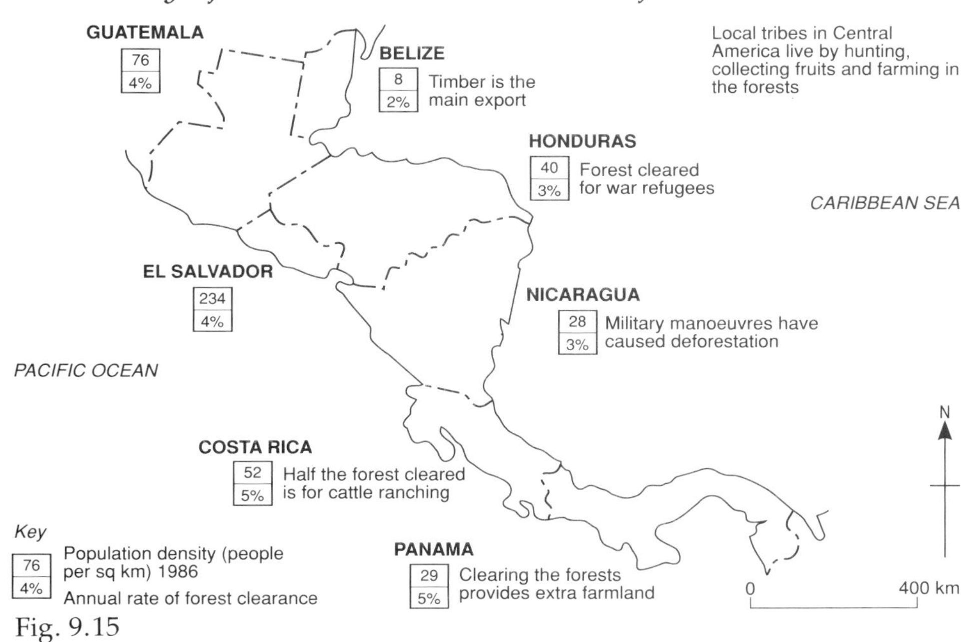

Fig. 9.15

1 Soil erosion rife in Panama and other Central American countries.
2 Panama Canal silting up.
3 Loss of soil fertility throughout Central America.
4 Growth of shanty towns in Central American cities.
5 Reduced rainfall in Panama.
6 Difficulty in flooding the locks in the Panama Canal.
7 Threatened extinction of the tapir, jaguar, ocelot.

Fig. 9.16

Solution 1: Debt for Nature
Costa Rica is protecting up to 15 per cent of its rainforest as National Parks. In return, it does not have to pay back the debts it owes some other countries. This gives it money for other developments.

Fig. 9.17

Solution 2: Agro-Forestry
The Inter-American Development Bank is funding agro-forestry in Costa Rica. Crops and trees are grown together. The trees provide fuel and keep the soil fertile, allowing higher yields from the crops beneath them.

Fig. 9.18

Fig. 9.19

Extension text

9E Effects of Clearing Rainforests:

On the Local Climate

Large scale deforestation of the tropical rainforests causes climatic changes. These in turn cause changes in the vegetation. In time, it is thought that the rainforests could become deserts. The diagram below shows how this might happen.

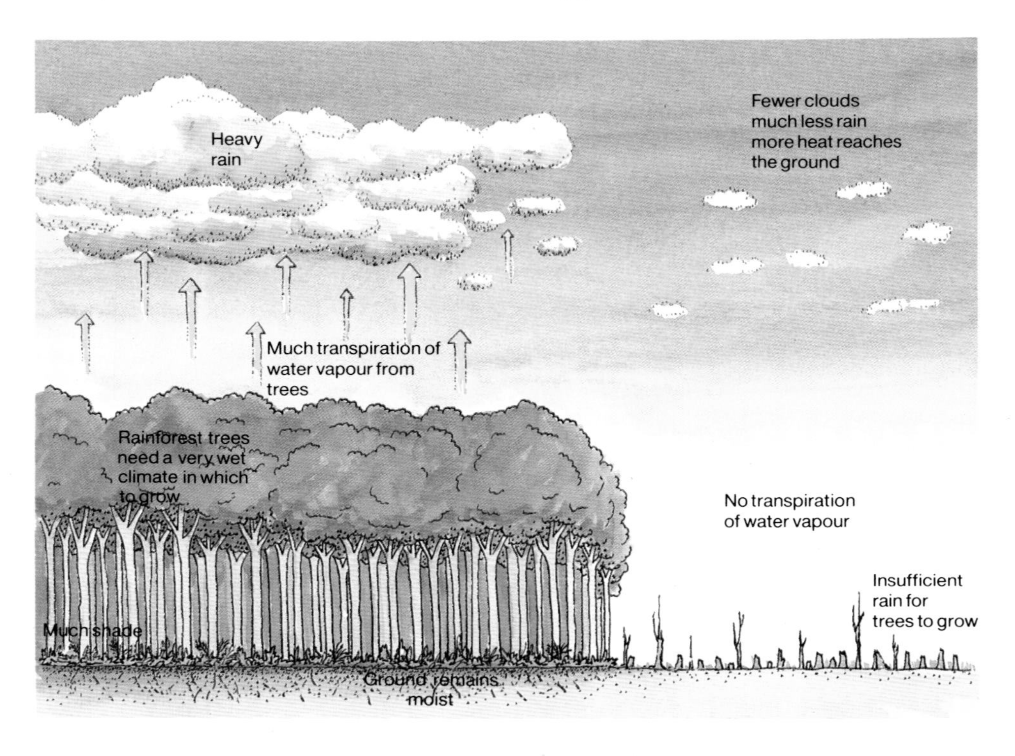

On the Greenhouse Effect

When large areas of tropical rainforest are cleared, temperatures around the world start to rise. This is called the **greenhouse effect.**

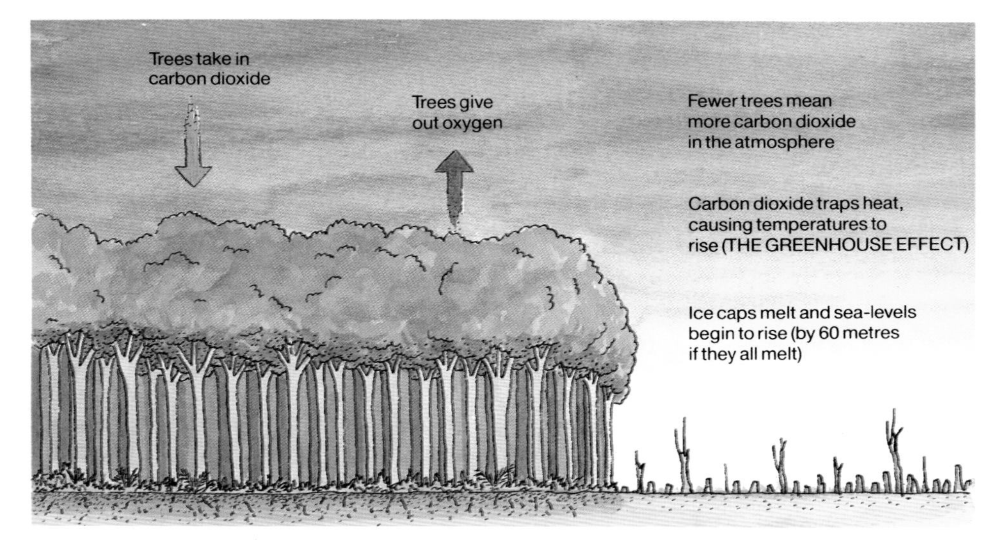

Questions

Look at 9E.

E1 Explain why rainfall is reduced when large areas of rainforest are cleared.

E2 In what way does the reduced rainfall and cloud cover affect the vegetation and temperature?

E3 Explain how clearing rainforests contributes to the 'greenhouse effect'.

E4 Explain how the 'greenhouse effect' affects the levels of seas and oceans.

Questions

Look at fig. 9.15.

C1 Compare the population density and the rates of deforestation within Central America.

'The purpose of clearing the Central American rainforests is to raise the standard of living of the local people.'

C2 Do you agree with the statement above? Give detailed reasons for your answer.

C3 How many of the events shown in fig. 9.16 are the result of deforestation? Give reasons for your answer.

Look at fig. 9.17 and 8.

C4 Which of the solutions do you think is better for the country of Costa Rica? Give reasons for your answer.

'Over half of Central America's remaining forests will be lost by the year 2000.'

'So what! It won't affect me in Scotland.'

C5 Explain in what way the last statement is not completely true.

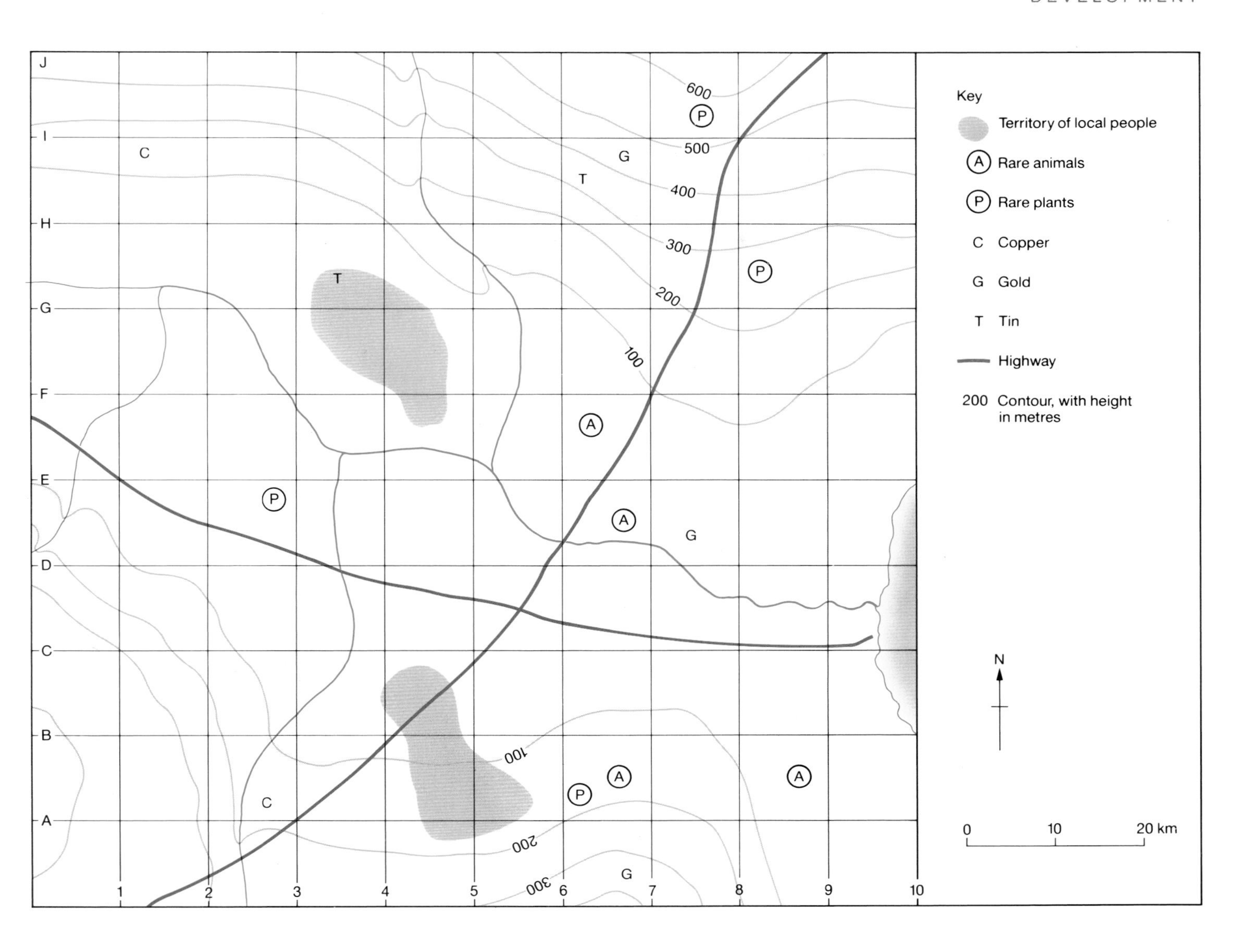

Core Groupwork

The Rainforest Game

The rainforest shown above belongs to one country. The country is going to develop its rainforest carefully. It has decided that

50 squares should be left as natural forest
25 made into farmland
10 for new settlements
10 for new industries (mines and factories)
5 for new reservoirs/power stations.

TASK 1 Working in groups, discuss which squares are best for farming, settlement, industry and reservoirs.
TASK 2 Draw a rough copy of the grid above on paper.
TASK 3 Each group takes it in turn to call out a square and suggest its land use (farming, forest, settlement, industry or reservoir) If correct the group scores 10 points. The teacher will explain which are the correct suggestions.

TASK 4 When a correct land use has been given, put the answer on your rough copy and make a note of the reason.
TASK 5 When all the squares have been correctly named, the group with the highest score is the winner.
TASK 6 Write a report explaining why the forest has been developed in that way.

UNIT 10 Using the Seas and Oceans

Core Text

Just like the tropical rainforests, our seas and oceans are being used more and more. They are used for fishing, mining, transport, recreation and dumping waste. As well as bringing benefits, using the seas and oceans can also bring problems.

10A Uses of the Oceans

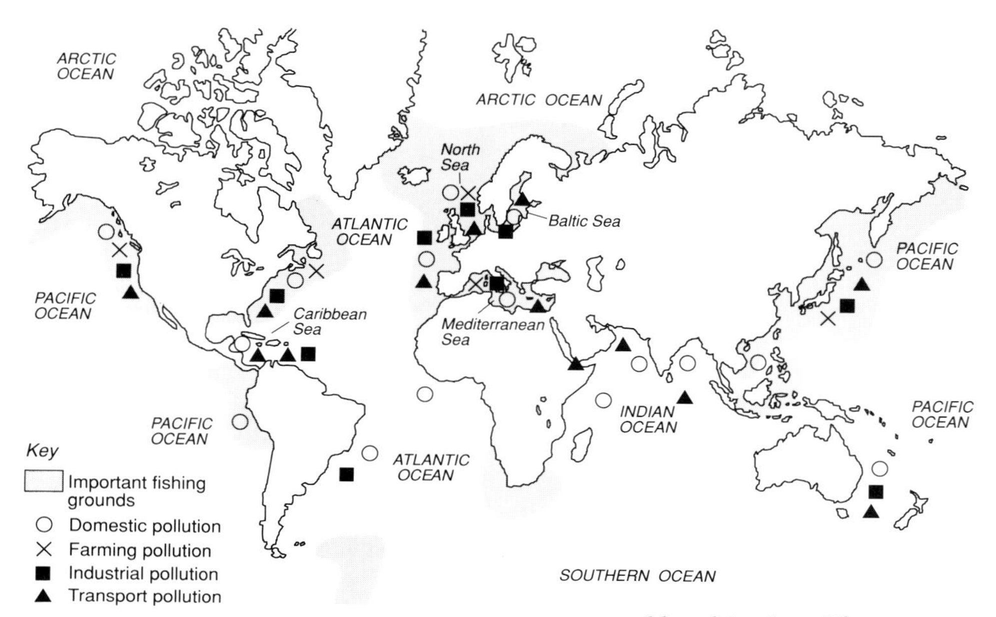

10B Ocean Pollution

Oceans can be polluted by farmers, factories and people along their coasts, and by shipping. They can also become polluted if the rivers that flow into them are polluted themselves.

Type of Pollution	Cause of Pollution	Effects of Pollution
FARMING	FERTILISERS PESTICIDES	Frothy sea water Sea slime Fish, plants and animals killed
INDUSTRIAL	DUMPING WASTE METALS AND CHEMICALS (eg CADMIUM, MERCURY) FALL-OUT FROM THE ATMOSPHERE	Fish killed Beaches polluted Humans eating fish are poisoned
DOMESTIC	DUMPING UNTREATED SEWAGE	Fish killed Beaches polluted Humans catch diseases eg typhoid, hepatitis, tetanus
TRANSPORT	OIL SPILLAGES	Beaches polluted Fish, animals and seabirds killed

10C Preventing Ocean Pollution

Farming – ban dangerous chemicals
Industrial – make laws to stop the dumping of waste into rivers and seas
Domestic – build sewage treatment plants
Transport – use aircraft to follow and monitor ships

10D Overfishing

Overfishing is when too many fish are caught. It means there will be fewer fish in the sea next year. The oceans are being overfished because:

1 Huge 'factory' ships are used, which can catch and process 1000 tonnes of fish every day
2 More countries with more boats are fishing all the oceans now
3 Some fishermen use nets with small mesh, which trap young fish.

10E Solving The Problem of Overfishing

1 Fishing limits – only one country allowed to fish within 320km of its coast
2 Make sure the fish nets have a wide mesh
3 Put a limit or **quota** on how many fish each fishing vessel can catch.

10F Whales in Danger

So many whales are now caught for their meat and oil that they are in danger of extinction (dying out).

	Numbers caught (in thousands)		
	1960	1974	1988
Fin Whale	32	2	<1
Sperm Whale	22	22	<1

Core Questions

Look at 10A.
1 Name four ways in which the world's oceans are used.
2 Which types of pollution affect the Atlantic Ocean?
3 Which seas and oceans suffer badly from industrial pollution?
4 In what way does the pollution of rivers affect the oceans?

Pollution froth in the sea

Look at 10B.
5 The froth in the photograph above is due to farming pollution. What causes farming pollution?
6 What are the effects of industrial pollution?

Look at 10C.
7 Describe one way in which domestic pollution can be prevented.

Look at 10D.
8 What is overfishing?

A factory ship

9 In what ways do factory ships cause overfishing?

Look at 10E.
10 Describe two ways of solving the problem of overfishing.

Questions

Case Study of the North Pacific Ocean

Look at fig. 10.1.
F1 In what ways is the North Pacific Ocean used?

Look at fig. 10.5.
F2 What problems are caused by using the Pacific Ocean?
F3 Describe how people in Japan were poisoned by mercury being dumped in the sea.

Look at fig. 10.4.
F4 Do you think that Japan's pollution laws are a good idea? Give reasons for your answer.

Look at figs. 10.2, 10.3 and 10.6.
F5 In 1989 there was a huge oil spill in the sea near Alaska. Which method or methods would best clear up the oil spill near Alaska?

Eskimos hunt whales for food

F7 The number of whales is falling rapidly. Who is more to blame – (a) the Eskimos (in the photograph above) or (b) the factory ships of the USSR and Japan (in the photograph to the left)? Give a reason for your answer.

Questions

Case Study of the Mediterranean Sea

Look at fig. 10.7.
G1 Describe the main uses of the Mediterranean Sea.

'The Mediterranean is one huge rubbish dump.'

Look at fig. 10.8.
G2 Which of the four solutions would be the best way of stopping outbreaks of typhoid? Give reasons for your answer.

Look at fig. 10.9.
G3 The statement above is exaggerated. Explain how it is exaggerated.

Look at fig. 10.10.
G4 Give one advantage and one disadvantage of the plan to reduce industrial pollution in the Mediterranean Sea.

Look at figs. 10.7 and 10.8.
G5 Do you think oil tankers should be banned from the Mediterranean Sea? Give reasons for your answer.

G6 In what way is the farmer's statement in the diagram above exaggerated?

Resources

Case Study of the North Pacific Ocean

The North Pacific Ocean is rich in fish and whales and the ocean-bed is rich in minerals. Many people and factories are found along its coast, especially in Japan. They dump their waste into the ocean. Oil tankers use the ocean to take crude oil away from Alaska.

Fig. 10.1

Effects of Alaskan Oil Disaster

1 Beaches polluted
2 Sea otters and sea lions drowned
3 Ducks and geese could not fly
4 Fish, whales and sea-birds poisoned by the oil.

Fig. 10.2

How To Clear Up an Oil Slick

1 Set fire to the oil
2 Use skimmers – boats which skim the oil off the surface of the water
3 Use booms – these are barriers which stop the oil from spreading
4 Use detergents.

Fig. 10.3

Laws in Japan make factories treat their waste
→ they should stop dangerous waste going into the sea
→ they may reduce industries' profits
→ industries may have to reduce their workforce
→ more jobs in factories making water treatment equipment.

Fig. 10.4

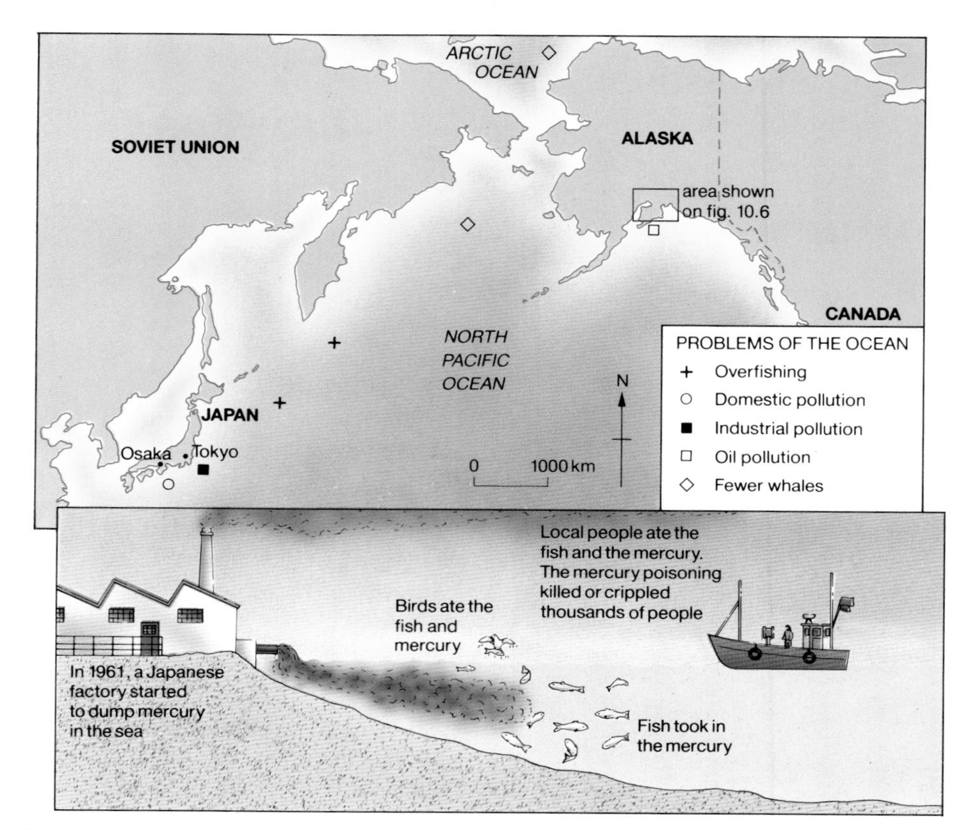

Fig. 10.5

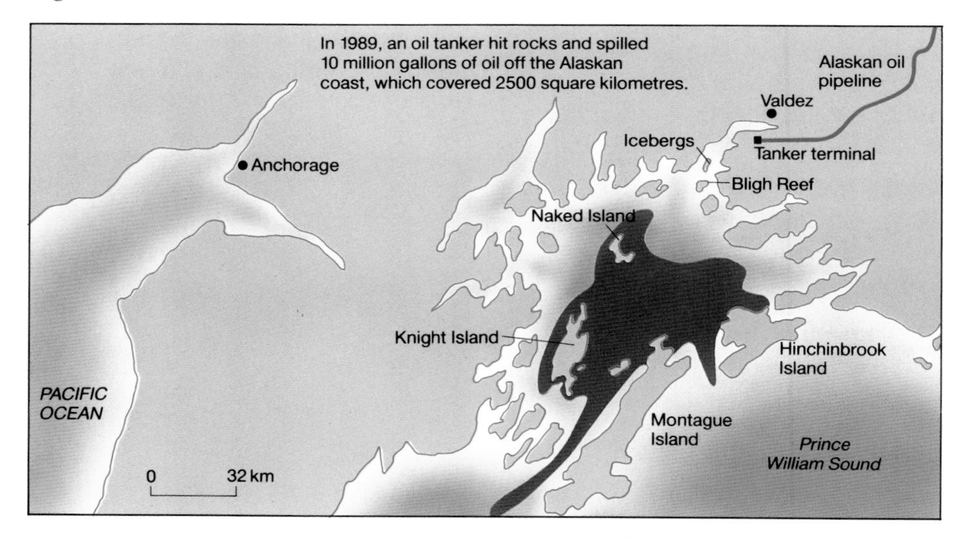

Fig. 10.6

Case Study of the Mediterranean Sea

Over 100 million people live along the coast of the Mediterranean Sea and the same number visit the area for holidays each year. 120 cities dump their domestic and industrial waste into the Mediterranean, while some factories use the sea water for cooling their equipment. Oil tankers use the Mediterranean as a quick route between the Middle East and Europe, while drilling for oil takes place near Italy. Small scale fishing goes on along the coasts but the sea is not well stocked.

Fig. 10.7

Preventing Domestic Pollution

1 Build expensive sewage treatment plants
2 Ban swimming near sewage outlets
3 Advise swimmers to be inoculated against diseases
4 Build pipelines to take sewage well out to sea.

Fig. 10.8

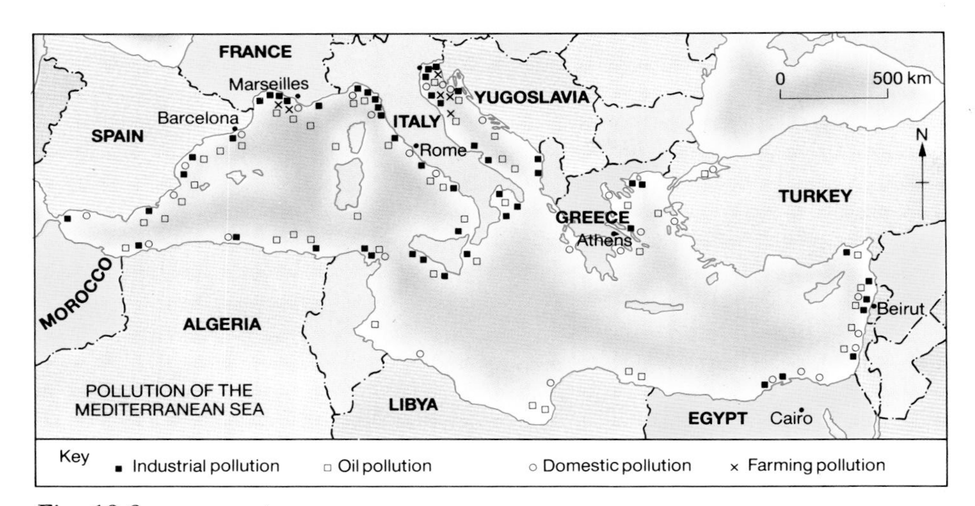

Fig. 10.9

Preventing Industrial Pollution

Seventeen out of the eighteen Mediterranean countries have agreed to ban the dumping of dangerous substances in the sea. Factories must install equipment to treat waste or dump it safely eg underground. Some factories now wish to move away from the Mediterranean.

Fig. 10.10

Case Study of the Southern Ocean

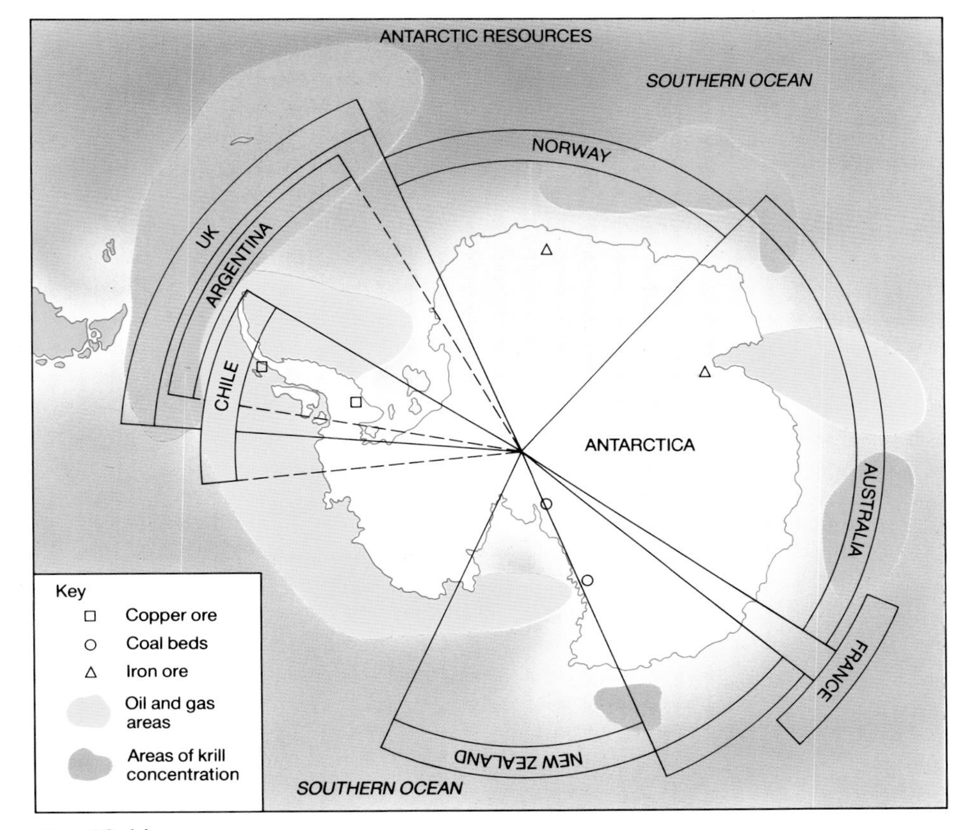

Fig. 10.11

Krill and plankton easily destroyed by pollution

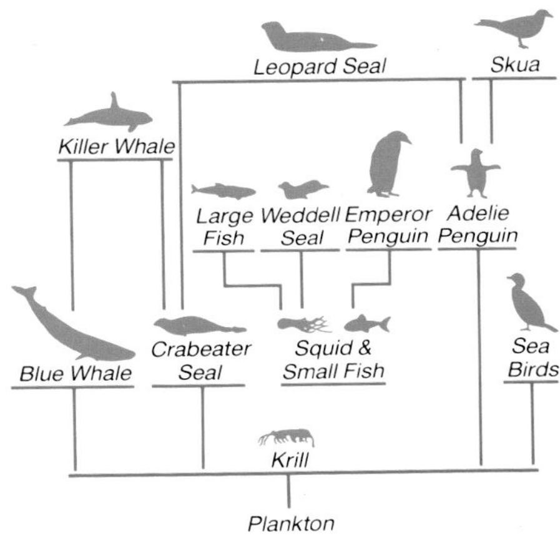

Despite the cold temperatures the Southern Ocean supports an abundance of life. Krill – small shrimp-like creatures – occur in large swarms. They are caught for animal feed. The largest mammal on Earth – the blue whale – lives here. So do the fin whale and the humpback whale.

Fig. 10.13

THE FUTURE OF THE ANTARCTIC REGION

World Park For research only	**Development Area** For the exploitation of all the region's resources
Conservation Area For research and tourism only	**Military Area** For the setting up of military bases

Fig. 10.12

Extension Text

10G Polluting the Seas and Oceans

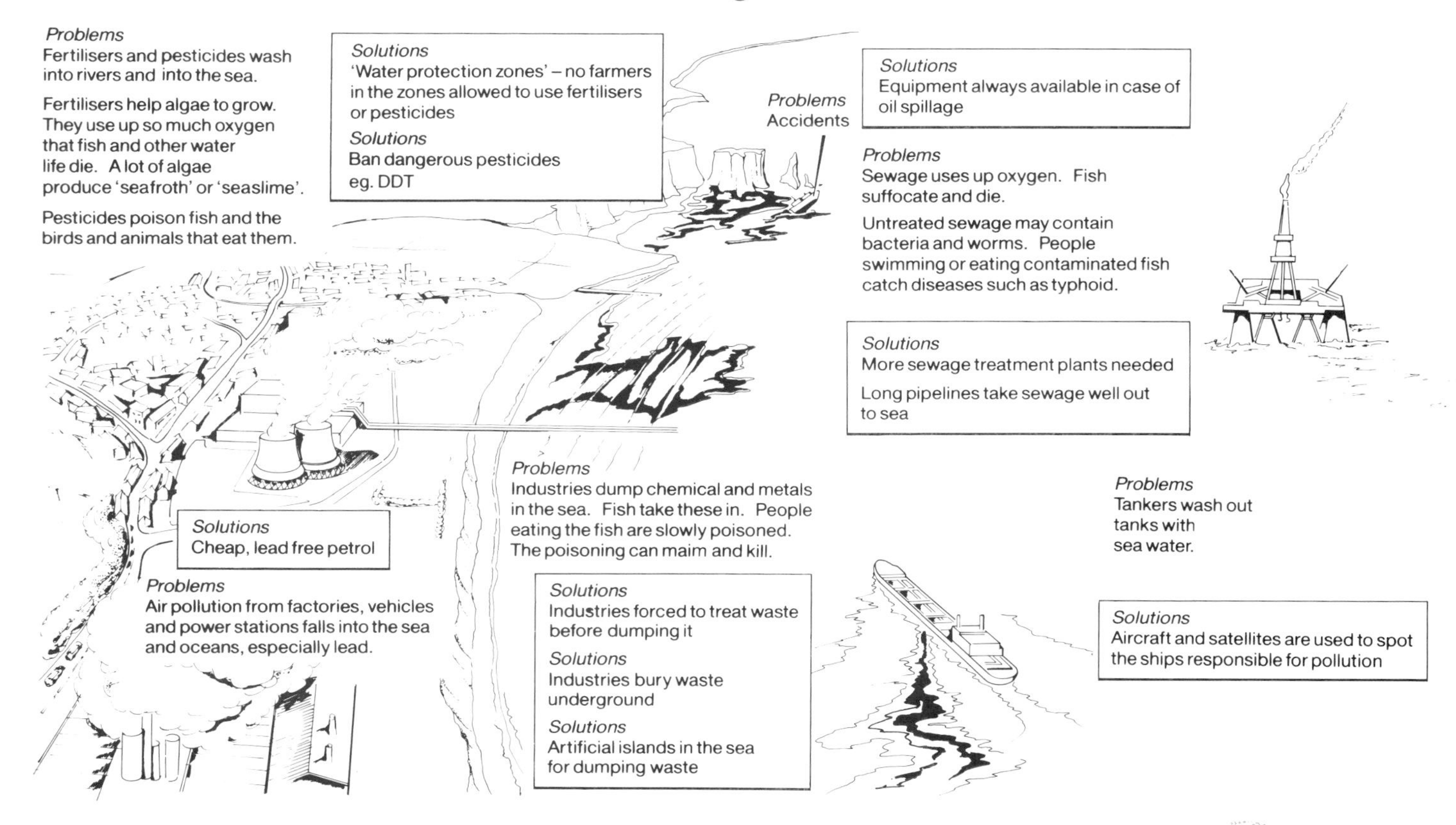

Questions

Look at 10G.

E1 Describe two causes of oil spills at sea.

E2 Explain how fertilizers affect the seas and oceans.

E3 Explain how people can be poisoned by industrial waste.

E4 Explain how sewage affects the oceans.

E5 Describe how oil spills at sea can be reduced.

E6 Describe how (*a*) industrial and (*b*) domestic pollution at sea can be reduced.

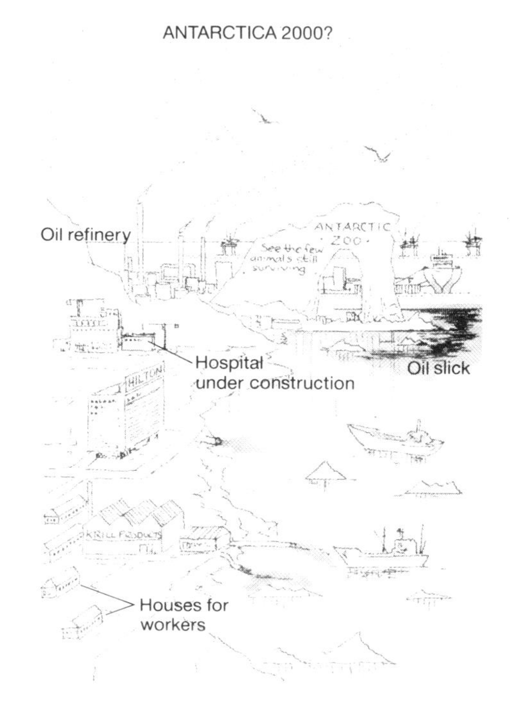

Questions

Look at figs. 10.11 and 10.13.

C1 What arguments might people put forward for and against the exploitation of the minerals of the Antarctic region?

C2 Describe the effects of over-exploiting the krill in the Southern Ocean.

Look at fig. 10.12.

C3 In what ways might the views of countries with and without claims to Antarctica differ over its future?

C4 What is the diagram to the left trying to show?

'Ocean pollution from tourism is easier to prevent than pollution from mineral exploitation'

C5 Do you agree with the above statement? Give reasons for your answer.

UNIT 11 Using the Land

Core Text

11A The World's Food

The land is one of the World's most important resources. Every year the land provides enough food to feed everyone in the world. But every year at least 40 million people die because of hunger, and only half of the world is well fed.

A WHERE DO ALL THE PEOPLE LIVE?

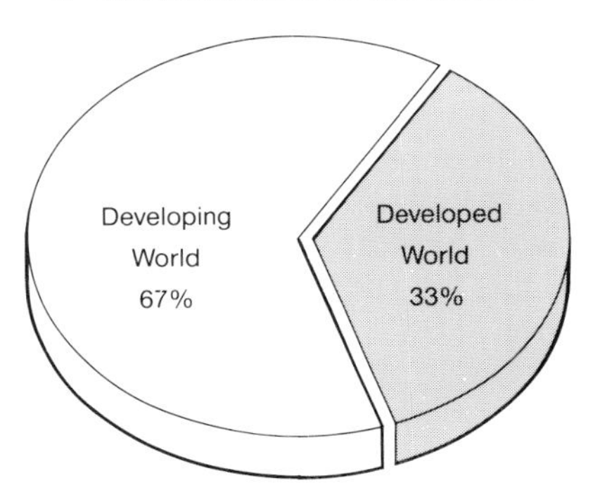

B WHO EATS THE WORLD'S FOOD?

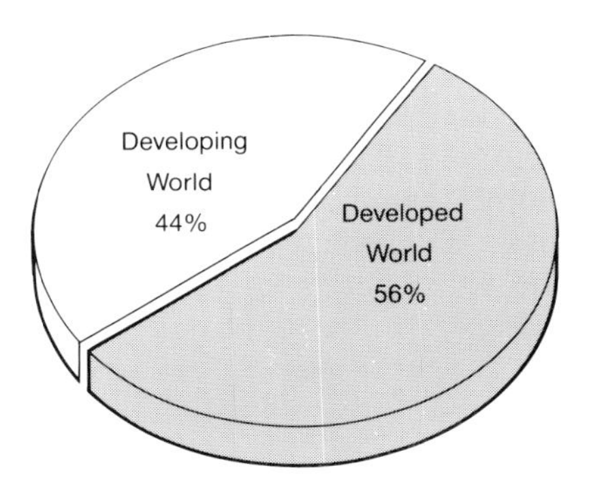

C WHO PRODUCES THE WORLD'S FOOD?

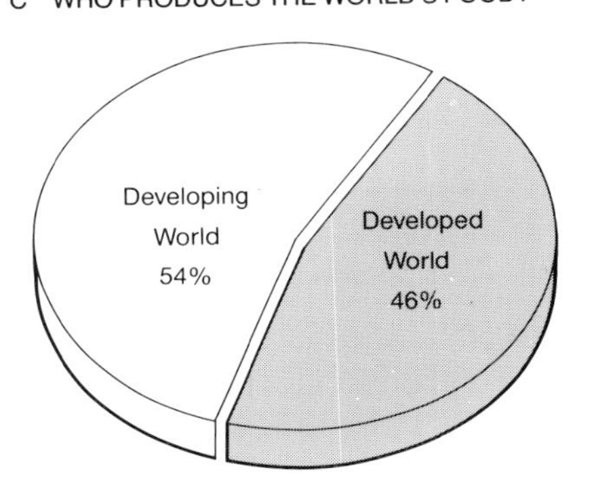

11B Farming Problems in the Developing World

The Developing World does not produce enough food for its people. There are many reasons why.

Problem

SOIL EROSION (soil taken away by wind or rain). The land turns to desert – this is called DESERTIFICATION

Caused by:

- DROUGHT (long periods with little rain). Crops and grass do not grow. The soil turns to dust and blows away.
- DEFORESTATION (cutting down trees). This is done for fuel and shelter. With no tree roots the soil is easily blown or washed away.
- OVERPOPULATION (too many people). Too many crops are grown and too many animals kept. Overgrazing makes the soil poor and easily eroded.

NATURAL DISASTERS

These are:

- FLOODS which wash away crops, animals, houses and people.
- CYCLONES which flatten crops and trees and block roads and railways.
- EARTHQUAKES and VOLCANOES which ruin farmland. They also kill people and animals, and destroy roads and railways.
- PESTS/DISEASES which eat one third of all the food grown, eg locusts, rats and mice.

SMALL FARMS. The population is growing rapidly, so less and less land is available per person. Also, rich people often own too much of the land.

This leads to:

SUBSISTENCE FARMING (farmers who grow only enough food for themselves). This means they have no money to improve their farms and produce more food.

11C Solving Farming Problems

Problem	Farming Solution
DROUGHT	**Irrigation** (putting extra water on the land) from rivers, wells and reservoirs. This stops soil erosion and allows more crops to be grown.
DEFORESTATION	**Plant more trees**. This stops soil erosion and makes the soil more fertile.
OVERPOPULATION	**Fertilisers** make crops grow better. **High yielding crops** produce a lot more food.
SOIL EROSION	**Shelter belts** of trees slow down the wind. **Terraces** (flat patches of land on hillsides) on sloping land slow down the run-off of rain.
PESTS/DISEASES	**Pesticides** allow more healthy crops to grow.
SMALL FARMS	**Share farmland more equally** so people have enough land to grow crops for sale. **Make more farmland** by irrigating deserts, terracing hillsides, draining marshes.

11D Food Production and Population

Each year the world grows more food than the year before. This is sometimes called the **Green Revolution**. But each year there are more people in the world than ever before.

	Increase in food in the last 20 years (%)	Increase in population in the last 20 years (%)
Western Europe	20	6
North America	35	12
South America	46	33
Africa	23	37
Asia	45	33

Core Questions

Look at 11B, 11C and 11D.
Drought; deforestation; overpopulation, desertification; subsistence; irrigation.
Which of the words above means:
(a) more food being produced,
(b) putting extra water on the land,
(c) long periods with little rain,
(d) cutting down trees,
(e) too many people,
(f) providing food only for yourself.

Questions

Look at 11A.
F1 Where do most people in the world live – in the Developed or Developing World?
F2 Who eats most of the World's food – the Developed or Developing World?
F3 Who produces most of the World's food – the Developed or Developing World?

Look at 11B.
F4 What two things take away soil?
F5 In what ways can soil erosion be stopped?
F6 In what ways can farms be made larger?
F7 Name two natural disasters that can ruin farmland.
F8 In what ways does irrigation help farmers?

High yielding rice being grown

Look at 11C.
F9 The photograph in the previous column shows the growing of high yielding rice. Explain how such crops help the farmers.

Look at 11D.
F10 In which region of the world has the population increased faster than food production?

Questions

Look at 11A.
G1 Compare pie graphs A and B.

G2 Why do many farmers in the Developing World only grow food for themselves?
G3 What causes desertification?

Look at 11D.
G5 Compare the increase in food production and population in Western Europe and Africa.

Crop irrigation

Look at 11C.
G4 The photograph above shows a farmer irrigating his land. Explain how irrigation helps the farmer.

Questions

Look at 11A.
C1 Compare pie graphs A, B and C. What conclusions can you draw from these graphs?

Look at 11B.
C2 Compare the importance of natural and human causes of farming problems in the Developing World.

C3 Describe what the table in 11D shows.

Look at 11B.
C4 The photograph above shows 'inter-cropping'. The farmer is growing a mixture of crops. Explain how this type of farming benefits the farmer.

Core Investigation

The next two pages give information on farming in the Sahel Region of Africa. Write a report on farming in the Sahel.
The aim of your investigation is to find out:
(*a*) The problems of farming in the Sahel.
(*b*) How the problems have affected the people.
(*c*) How some of the problems have been solved.
Your answer should be in four parts:
1 Aim
2 Method
3 Analysis
4 Conclusion

Use other books and atlases to get more information. Look at page 16 to find out how to write up your report.

Case Study of the Sahel Region, Africa

The Sahel is an area of Africa, just south of the Sahara Desert. It gets between 100 and 500 mm of rain each year, on average. Because there is little rain, most farmers are nomadic herders – they move around with their cattle, sheep and goats in search of water and grazing. Some farmers can grow crops such as millet and sorghum. Since 1970, the Sahel has suffered from **desertification** – much of the land is turning into desert.

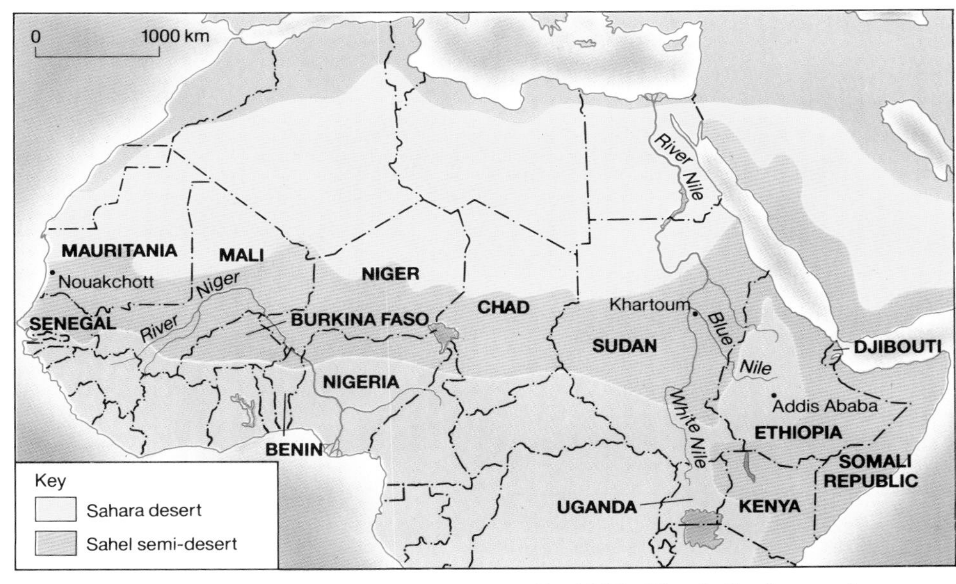

THE PROBLEMS

Rainfall in Nouakchott (Mauritania)

150 mm of rain should fall here each year. This would be enough for grass for animals to graze.

Year	Rainfall	Year	Rainfall
1968	60mm	1983	10mm
1969	140	1984	5
1970	45	1985	35
1971	25	1986	60

Dust Storms in Nouakchott (Mauritania)

Year	Days of Duststorms	Year	Days of Duststorms
1968	1	1982	52
1969	3	1983	85
1970	7	1984	80
1971	16	1985	82
1972	30	1986	80

In Ethiopia 45 per cent of the farms are less than 1 hectare in size

People and animals crowd around the few wells, eating and trampling the grass

The Sahel: People and Livestock

	Population (millions)	Livestock (millions)
1950	15	30
1970	25	66
1990	43	69

Some farmers in the Sahel grow only cash crops. These make the soil infertile. When the rains fail, they have no money or food. In Mali, food production fell rapidly in the 1970s but cotton production rose by 400 per cent.

In 1985, Chad, Sudan and Ethiopia were at war. Over half of Ethiopia's money was spent on arms.

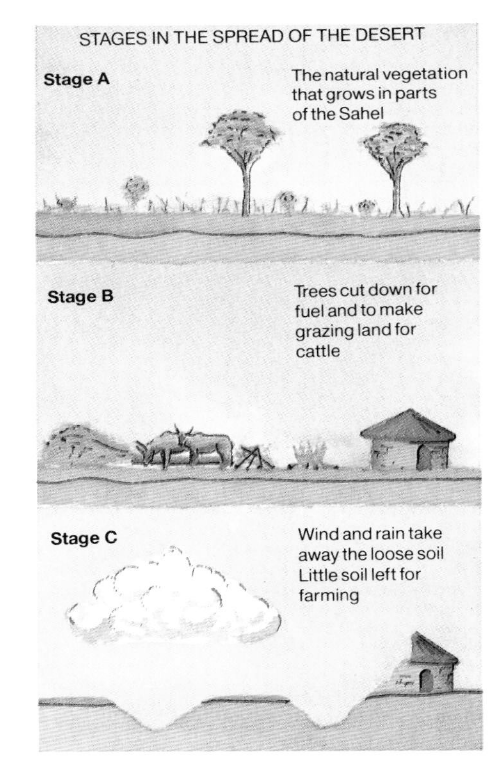

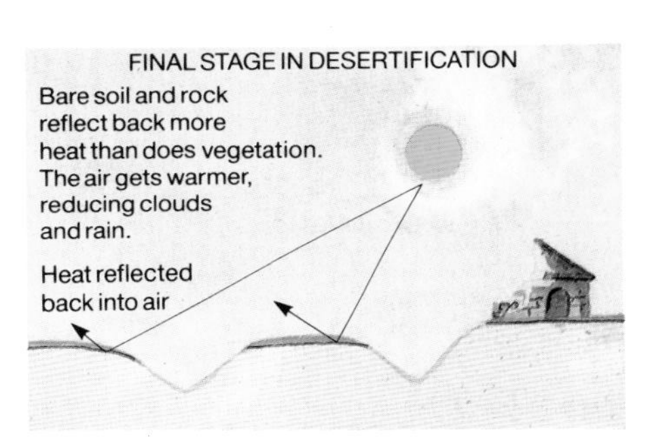

THE EFFECTS

Between 1968 and 1974 ¼ million people in the Sahel died of hunger; 20 million sheep, cattle and goats died; crop farmers had to eat their seeds, so they had none left to sow next year; herders lost all their animals.

Between 1984 and 1986 another 2 million people died from the drought in the Sahel.

Population of Nouakchott (Mauritania)	
1960	20 000
1980	200 000
1990	500 000

Migration of People	
65 000	from Ethiopia to Somalia
700 000	from Ethiopia to Sudan
120 000	from Chad to Sudan
300 000	to the city of Khartoum, Sudan
300 000	to the city of Addis Ababa, Ethiopia.

POVERTY CYCLE

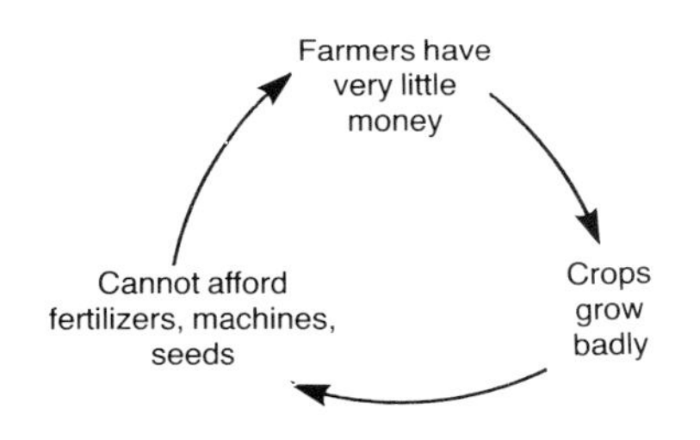

DISEASE CYCLE

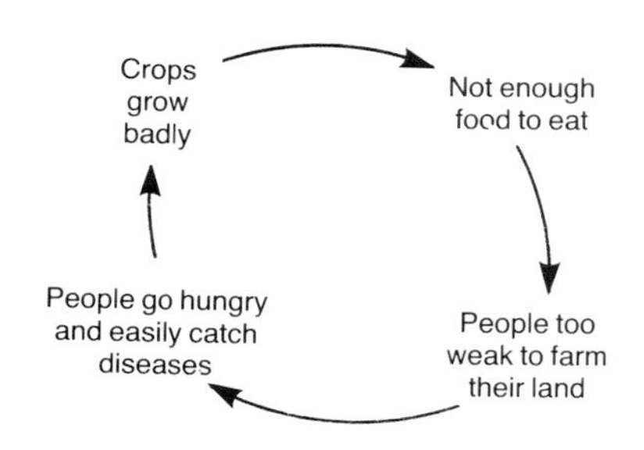

Famine in Ethiopia

THE SOLUTIONS

The Gezira Scheme in the Sudan irrigates land from the Blue and White Niles using canals and ditches. Two harvests a year are now possible. Cotton is grown for export. Wheat, sorghum and vegetables are grown for food.

There are 200 000 kilometres of terraces in Ethiopia. They help trap water, stopping it from washing the soil away

Spraying pesticides on grasshoppers and locusts in 1986 saved 90% of the crops in the affected areas.

Putting fertilisers on cereals in Ethiopia increased the yields from 750 kg/ha to 1050 kg/ha.

An irrigation scheme in Senegal grew aubergines and mangoes for people in Europe, while people in Senegal were suffering from hunger.

Ferns, planted in rice fields in Burkina Faso, increased soil fertility. Yields rose by 42%. The ferns also provide fodder and keep down weeds.

Agro-forestry in the Sahel has increased yields of millet and groundnuts from 500 kg/ha to 900 kg/ha.

A tree nursery in Ethiopia. Trees stop the soil blowing away and make the soil more fertile. Niger planted 1 million trees in 1986

A biogas machine making heat and fertilizer from animal dung

UNIT 12 Improving People's Health

Core Text

12A The Importance of Health

The forests, oceans and farmland are all important resources, but the most important resources of all are people. It is people who use the forests, oceans and farmland. And if we are to continue to make better use of these resources, we need to be healthy. Unfortunately, most of the World's people suffer from at least one disease. This not only affects their ability to enjoy life, but also their ability to work.

12B Diseases Caused by Lack of Food

Lack of food makes people weak. They cannot work hard and they easily catch other diseases. Over 2000 million people suffer from lack of food. To be healthy, adults need to eat each day:

(a) 2500 calories from carbohydrates in food such as bread, potatoes and sugar.
(b) 65 gm protein in meat, eggs, milk.
(c) Vitamins and (d) minerals from fruit, vegetables, dairy produce.

People are **undernourished** or starving when they do not eat enough calories.

People are **malnourished** when they do not eat enough protein, vitamins and minerals. They do not have a balanced diet.

12C Diseases Spread by Polluted Water

Three-quarters of all the diseases in the Developing World are spread by polluted water. Every year 25 million people die from these diseases.

CHOLERA –	Kills 10 million people every year
SNAIL FEVER –	Affects 300 million people and 1 million die each year – people too weak to work
GUINEA WORM DISEASE –	Affects 20 million people each year. Many are crippled – people too weak to work
DIARRHOEA –	Kills many children

HOW TO PREVENT THESE DISEASES
1. Clean water
2. Treat sewage
3. Vaccination

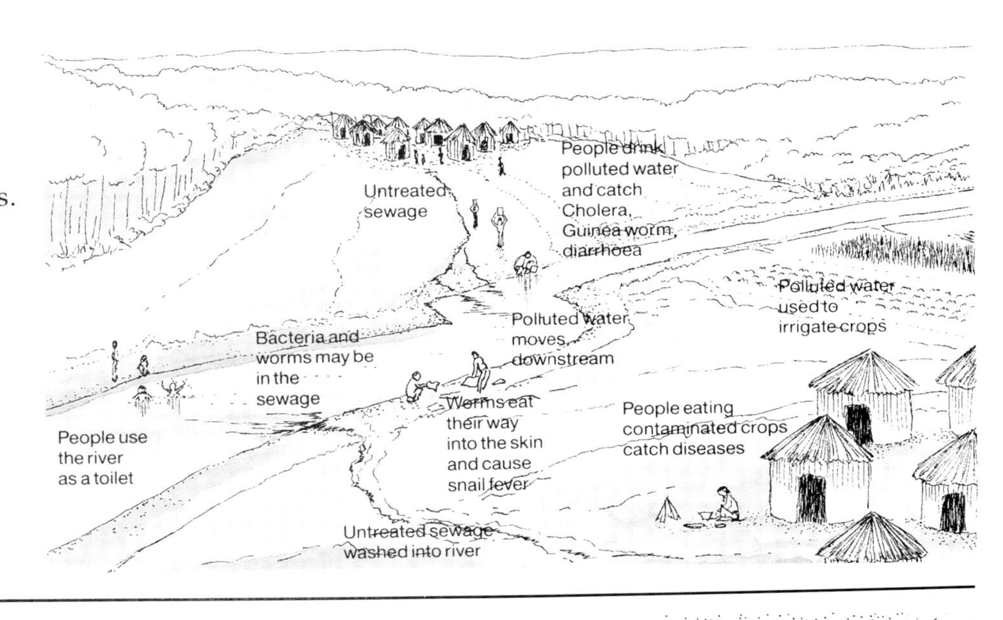

12D Diseases Spread by Flies and Mosquitoes

Many flies and mosquitoes are found in hot areas of the World. This is where most of the Developing Countries are found.

RIVER BLINDNESS –	Causes blindness and kills 200,000 people each year
MALARIA –	Affects 400 million people and kills 8 million each year. Causes fevers and weakens people
SLEEPING SICKNESS –	Affects 200 million people. People are too weak to work

HOW TO PREVENT THESE DISEASES
1. Use chemicals to kill the flies and mosquitoes
2. Destroy breeding grounds of the flies and mosquitoes
3. Vaccination

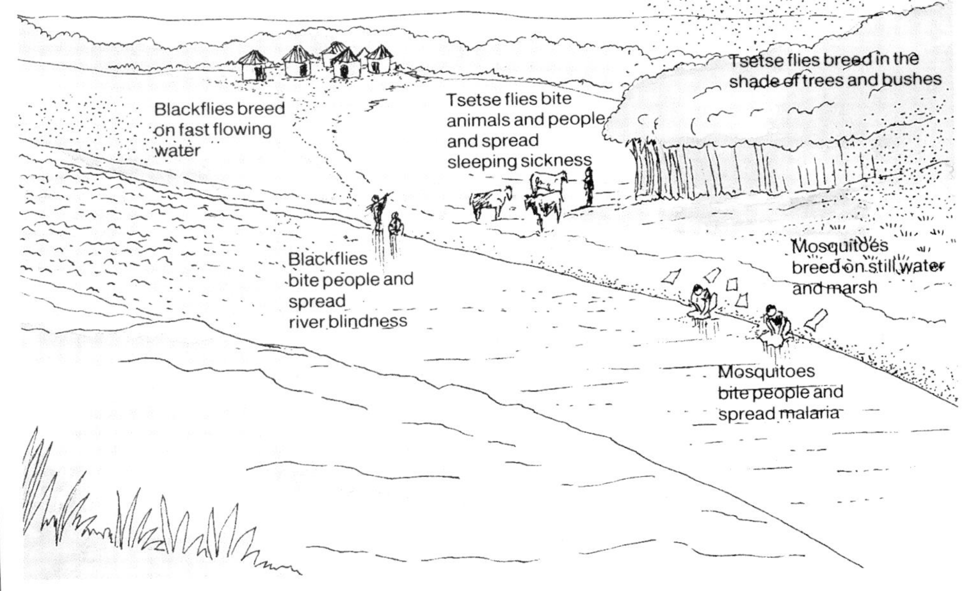

Core Questions

Look at 12B.

1 What four things do people need to eat to be healthy?

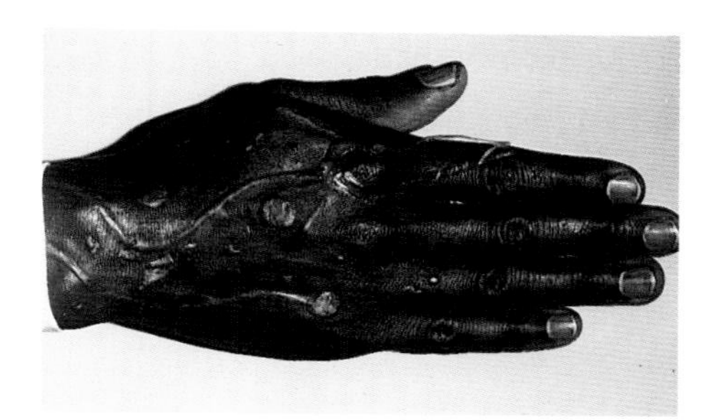

Guinea worm breaking through the skin

Look at 12C.

2 Explain how people catch guinea worm disease.

3 Name two other diseases spread by polluted water.

4 Explain how water becomes polluted with bacteria and worms.

5 Name two ways of preventing disease spread by polluted water.

Look at 12D.

6 What disease do blackflies spread?

7 How does sleeping sickness affect people?

8 How do people catch malaria?

Spraying rivers with chemicals to reduce river-blindness

9 How does the spraying of rivers reduce river blindness?

10 Explain how clearing trees can reduce sleeping sickness.

Questions

Case Study of Bangladesh

Look at fig. 12.4.

F1 Do you think people in Bangladesh catch diseases from the River Ganges? Give reasons for your answer.

F2 In which place on the map, A or B, are people more likely to have malaria? Give a reason for your answer.

F3 What does fig. 12.1. show?

'People in Bangladesh don't have enough food because they are too lazy to work their land properly.'

Look at fig. 12.2.

F4 The statement above is biased. Explain how it is biased.

Water pumps in Bangladesh

Look at fig. 12.5.

F5 Describe how pumps, shown in the photograph above, help the people of Bangladesh.

Look at fig. 12.6.

F6 Which method, 1, 2 or 3, should Bangladesh use to get rid of malaria? Give a reason for your answer.

Look at fig. 12.3.

F7 Which of the following best explains why so many children die in Bangladesh?

(*a*) Their diseases cannot be treated

(*b*) There are too few doctors

(*c*) The diseases are too expensive to treat.

Give a reason for your answer.

Questions

Case Study of Burkina Faso

Look at fig. 12.7.

G1 Describe the distribution of major diseases in Burkina Faso.

'People in Burkina Faso should not live near rivers'

G2 Give one argument for and one against the statement above.

Look at figs. 12.8 and 12.9.

G3 Which disease – malaria or sleeping sickness – will most affect the amount of food produced in Burkina Faso? Why?

'The health of the people of Burkina Faso is steadily improving'

Look at fig. 12.10.

G4 Do you agree with the statement above? Give reasons for your answer.

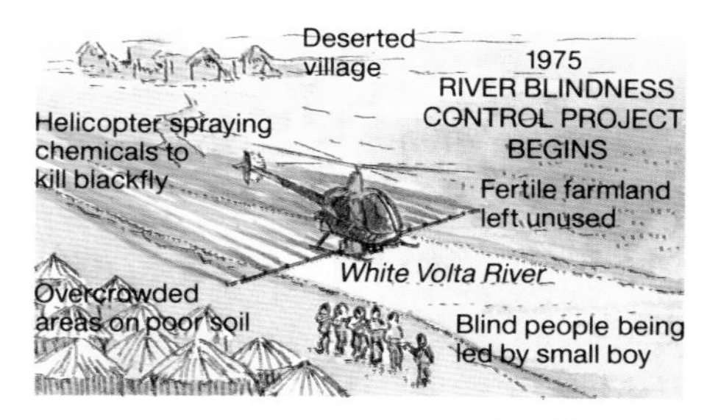

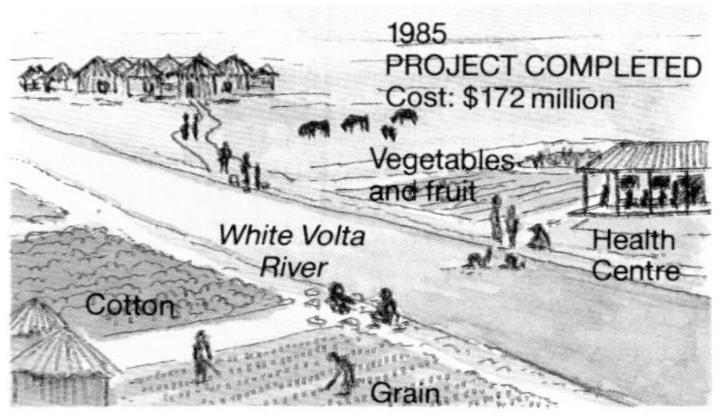

Look at the diagrams above.

G5 Describe the effects of the river blindness control project on the people of the White Volta region.

Look at fig. 12.11.

G6 Which improvement – the pump or the latrine – would most reduce disease in Burkina Faso? Give reasons for your answer.

Resources

Case Study of Bangladesh

	Amount needed to be healthy	Amount eaten in Bangladesh (1985)
Calories	2500	1812
Protein (gm)	65	41

Fig. 12.1

Common Diseases in Bangladesh

Disease	Effects
Cholera	Impossible to work; kills many people
Malaria	Kills many children; people unable to work
Starvation and malnutrition	People too weak to work

Fig. 12.2

Causes of Child Deaths in Bangladesh

Disease	Cost of treatment
Tetanus	15p per child
Diarrhoea	3p per child
Measles	8p per child

Fig. 12.3

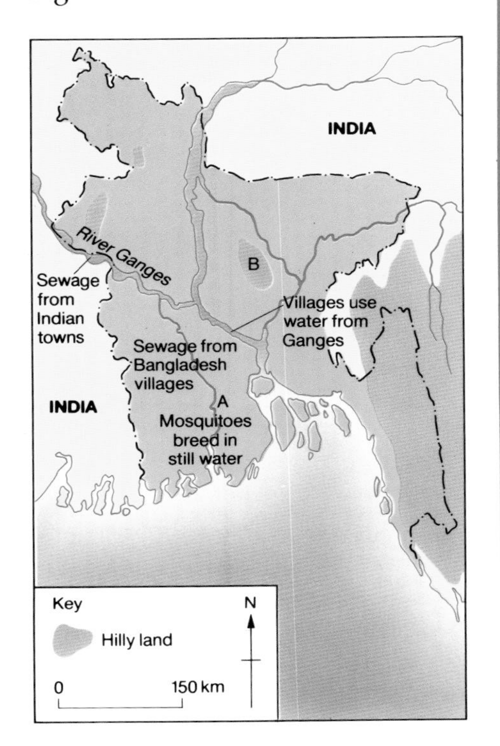

Fig. 12.4

Effects of New Water Pumps

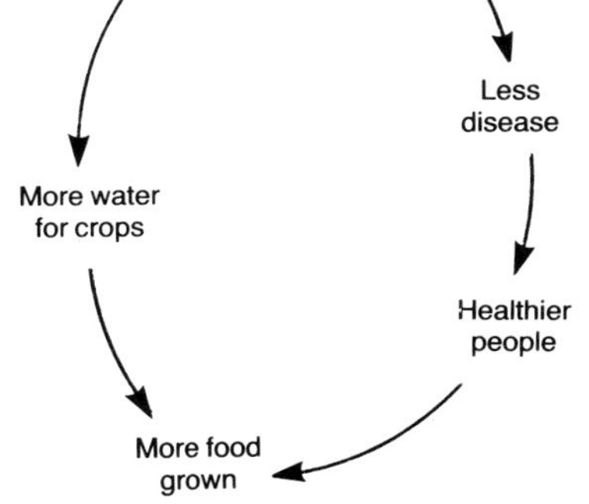

Fig. 12.5

How to Prevent Malaria

1 Drain areas of still water – but still water is needed to grow rice
2 Spray mosquitoes with insecticides – very expensive
3 Give drugs – stops people dying, but it does not kill the mosquito.

Fig. 12.6

Case Study of Burkina Faso

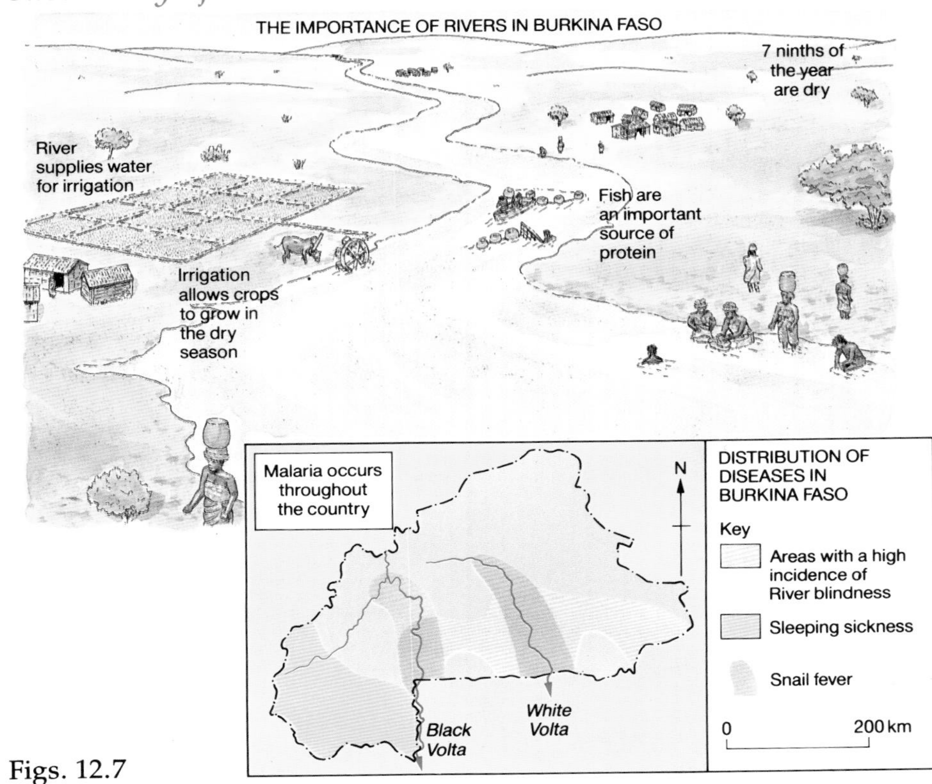

Figs. 12.7

Effects of Malaria
1 People move away from river valleys
2 Adults get fevers and are unable to work
3 People easily catch other diseases

Fig. 12.8

Effects of Sleeping Sickness
1 Makes adults too weak to work
2 Cattle die
3 People move away from forested areas

Fig. 12.9

Burkina Faso	1960	1980
Infant mortality	252‰	160‰
Life expectancy	37	44
Calories per day	2020	1922
Death rate	28‰	22‰

Fig. 12.10

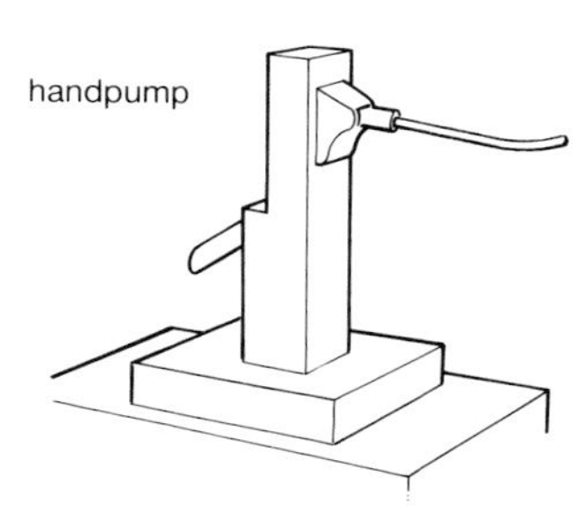

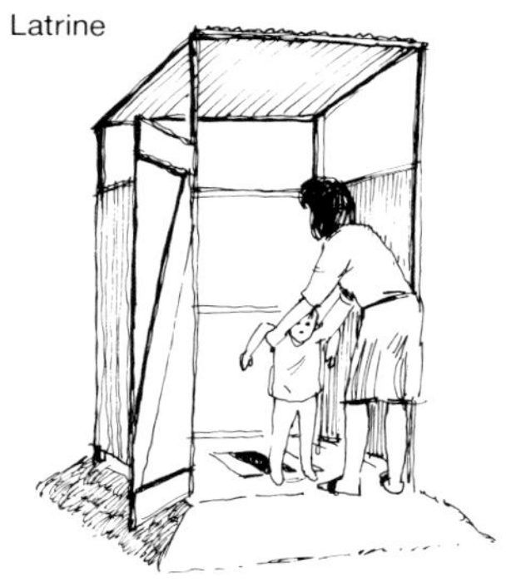

Fig. 12.11

Case Study of Nigeria

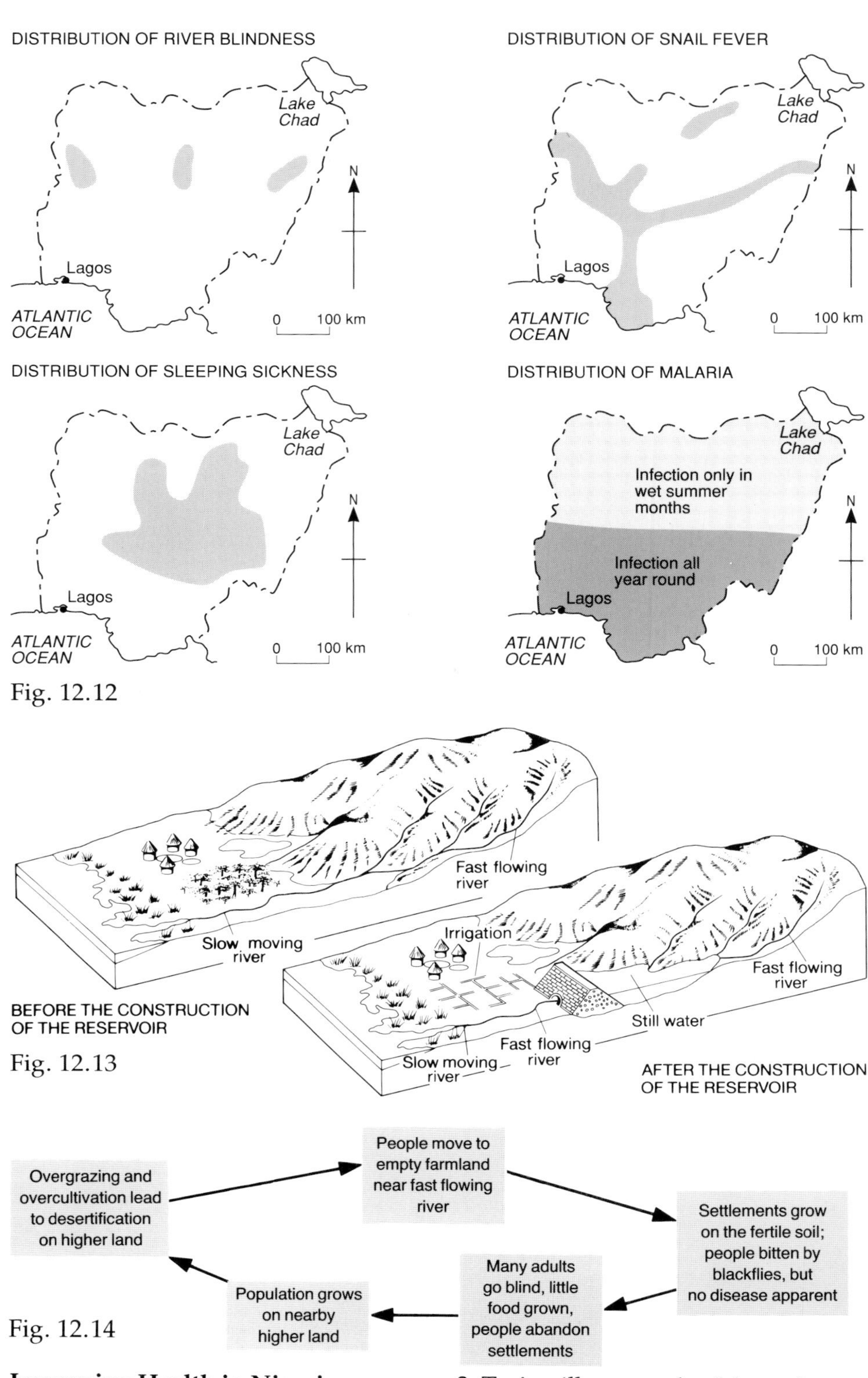

Fig. 12.12

Fig. 12.13

Fig. 12.14

Improving Health in Nigeria
1 Install handpumps which bring clean water up from underground.
2 Install latrines, for safe disposal of sewage.
3 Train villagers as health workers to explain to others how disease is spread and how it can be prevented.

Fig. 12.15

Extension Text

12E Contrasts in World Health

Developing World
Most people suffer from disease
Many children die of disease
Most people suffer from hunger
Infectious diseases are very common
Life expectancy is 60 years

Developed World
Most people are healthy
Few children are unhealthy
Little hunger
Non-infectious diseases are most common
Life expectancy is 75 years

Diseases affect the Developing Countries much more severely than the Developed Countries. Not only do the diseases kill, they can also blind, cripple and weaken whole communities of people. This drastically reduces food production which, in turn, leads to starvation and malnutrition. In addition, diseases such as malaria and river blindness have forced people to move away from fertile areas beside rivers. This causes overcrowding and the eventual desertification of neighbouring land. Disease is, therefore, the single biggest factor explaining the difference in wealth between the Developing and Developed Worlds.

12F Factors in World Health

DEVELOPING WORLD		DEVELOPED WORLD	
Population per doctor	15000		500
People with clean water	54%	People with clean water	96%
People with proper sanitation	35%	People with proper sanitation	94%

Most diseases in the Developing World could be prevented if only there were enough doctors, medicines, safe water supplies and sanitation. But to afford these improvements, the Developing Countries need to produce more wealth. This, of course, is difficult to do when so many people suffer from disease.

12G Improvements in World Health

The table below shows figures for Developing Countries in Column A and for Developed Countries in Column B.

Change (1970-85)	A	B
Life expectancy	+15%	+3%
People with safe water	+34%	+4%
People with proper sanitation	+76%	+7%

Because of their poverty, methods of improving health in the Developing World must be cheap as well as effective. The most successful methods involve the local community in the improvements. This is called **primary health care**. It can take many forms. Villagers can be trained as health workers to treat common diseases and educate people on how to avoid diseases. Others can make simple handpumps which provide a whole village with clean water. Others can make cheap latrines out of local materials, which reduce pollution.

Questions

Look at 12E.
E1 How does disease affect food production in the Developing World.

Look at 12F.
E2 Explain why disease is more common in the Developing World than the Developed World.
E3 Explain why it is difficult for Developing Countries to improve the health of their people.

Look at 12G.
E4 Describe what is meant by 'primary health care'.

Questions

Look at fig. 12.12.
C1 Compare the distribution of diseases in Nigeria.

Look at fig. 12.14.
C2 Describe the relationship between river blindness and food production in North Nigeria.

Look at figs. 12.13.
C3 Do you think that new irrigation schemes in Nigeria improve the health of the local people? Explain your answer fully.
C4 People in Nigeria disagree over which of the three methods, shown in fig. 12.15, would most reduce disease in each village. Describe the different arguments that people would put forward.

'Nigeria should get rid of all its tsetse flies, blackflies and mosquitoes. Then its people will never again have to suffer the diseases they spread'
C5 The statement above is exaggerated. Explain how it is exaggerated.

Core Text

13A Imports and Exports

Countries can improve their standard of living by selling more goods abroad (**exports**). This gives them money to buy goods from other countries (**imports**).

VALUE OF EXPORTS − COST OF IMPORTS = BALANCE OF TRADE

The difference between the value of imports and exports is called the **balance of trade**. Most Developed Countries get more money from their exports than they have to pay for their imports. But most Developing Countries do not sell enough exports to pay for all the imports they need to buy.

13B Patterns of International Trade

Manufactured goods are those which have been made eg steel, computers. **Primary goods** are foodstuffs and resources which have not been made eg coal, wheat, wood. They provide the raw materials for making manufactured goods.

Developed Countries ('**The North**') have many factories making manufactured goods, but they need primary goods as raw materials in their factories.

Developing Countries ('**The South**') have few factories so they import manufactured goods. They sell primary goods to pay for them.

The Developed and Developing Countries are **interdependent** – they need each other.

	Developed Countries' Trade	Developing Countries' Trade
EXPORTS	Mostly manufactured goods	Mostly primary goods
IMPORTS	Mostly primary goods	Mostly manufactured goods
AMOUNT OF TRADE	A large amount	Little
NUMBER OF EXPORTS	Many	Very few

13C Problems of North

13D Barriers to International Trade

Many countries make it difficult for others to exort their goods to them by (a) putting a **quota**, or limit, on the amount of goods they import and (b) having **tariffs**, or taxes, which are added to the price of imported goods.

13E Trading Alliances

Some countries have grouped together as **trading alliances**. They allow each country in their group to import as many goods as they wish. There are no quotas or tariffs between them. One example is the European Community.

Some countries have grouped together to export one product. Instead of trying to sell their product more cheaply than the other countries, they agree to sell them at the same high price. One example is the Organization of Petroleum Exporting Countries (OPEC).

Core Questions

Look at 13A.
1 What are imports and exports?
2 What is a country's 'balance of trade'?

Look at 13B.
3 For each of the goods below, write down whether they are manufactured or primary:
(*a*) steel (*b*) wheat (*c*) machines
(*d*) tables (*e*) oil (*f*) fish
4 How are the exports of Developed and Developing Countries different?

5 The cartoon above shows that the countries of 'the North' and 'the South' are interdependent. How does the cartoon show this?

Look at 13C.
6 In what ways do the prices of primary goods change?
7 Why is this bad for the Developing Countries?
8 In what way do the prices of manufactured goods change?
9 Why do Developing Countries have to borrow money?

'It is a limit on the amount of goods a country can import'
'It is a tax on imports'

Look at 13D.
10 Which of the statements above describes (*a*) a tariff (*b*) a quota?

Questions

Case Study of Dominica, West Indies

Look at fig. 13.1.
F1 Compare Dominica's imports and exports.

'Dominica exports mostly primary goods and imports mostly manufactured goods'.

Look at fig. 13.3.
F2 Do you agree with the statement above? Give reasons for your answer.

Look at fig. 13.4.
F3 Do you think Dominica was richer in 1982 than in 1970? Give a reason for your answer.

F4 20 per cent of Panama's exports are bananas.
50 per cent of Dominica's exports are bananas.
Which country – Panama or Dominica – will be worse affected by a drop in the price of bananas? Give a reason for your answer.

Bananas account for 50% of Dominica's exports.

Look at the photograph above.
F5 Do you think that Hurricane David affected Dominica's exports and imports? Give reasons for your answer.

Look at fig. 13.5.
F6 Do you think it is good or bad for Dominica that she sells all of her bananas to one company? Give reasons for your answer.

Look at fig. 13.6.
F7 Which of the three solutions is best for Dominica? Give reasons for your answer.

Questions

Case Study of Zambia, Africa

Look at fig. 13.9.
G1 Describe the main differences between Zambia's imports and exports.

Look at fig. 13.8.
G2 Describe Zambia's balance of trade between 1964 and 1978.
G3 In 1965, Zambia's started a 10 year programme to build more schools, roads and other services.
Was this the best year to start a 10 year programme? Give reasons for your answer.

Look at fig. 13.10
G4 What were the main differences between the changes in price of copper and agricultural vehicles between 1960 and 1986?
G5 Zambia uses its money from copper to buy vehicles. Would it have been able to buy more or fewer vehicles in 1960? Give reasons.

(A) Plastic piping replaces copper piping
(B) Railway from copper mine to port blown up
(C) New copper mines open in Australia, Canada and Mexico

G6 Which of the three events above (A, B or C) would do the most harm to Zambia? Explain what harm would be done.

Look at fig. 13.11.
G7 Which of the three solutions is best for Zambia's trade problem? Give one advantage and one disadvantage of the solution you have chosen.

Resources

Case Study of Dominica, West Indies

Dominica is a tiny island, about the same size as the Isle of Skye. It is in the Windward Group of the West Indies.

Imports (1988) £38 million
Exports (1988) £28 million

Fig. 13.1

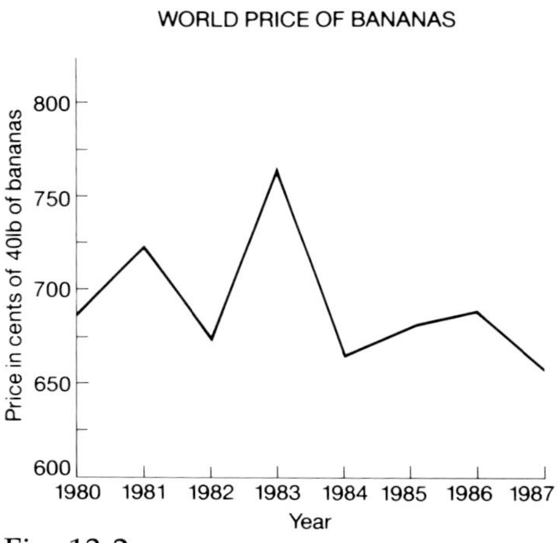

WORLD PRICE OF BANANAS

Price in cents of 40lb of bananas: 600, 650, 700, 750, 800

Year: 1980, 1981, 1982, 1983, 1984, 1985, 1986, 1987

Fig. 13.2

Main Exports	Main Imports
Bananas (50%)	Machinery
Citrus fruits	Chemicals
Fruit juices	Food

Fig. 13.3

Price of bananas higher in 1982 than in 1970.

Price of fertilizers doubled between 1970 and 1982.

In 1970, 1 tractor cost the same as 11 tonnes of bananas.

In 1982, 1 tractor cost the same as 25 tonnes of bananas.

Fig. 13.4

Dominica's bananas are all bought by one multinational company. This company brings advantages and disadvantages to Dominica.

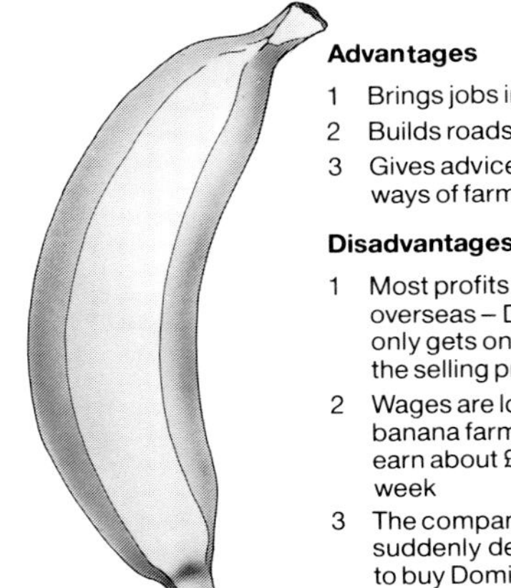

Advantages
1 Brings jobs in Dominica
2 Builds roads and ports
3 Gives advice on better ways of farming

Disadvantages
1 Most profits go overseas – Dominica only gets one third of the selling price
2 Wages are low – banana farmers earn about £5 per week
3 The company may suddenly decide not to buy Dominica's bananas

Fig. 13.5

Solution to Dominica's Trade Problem

A Join the Union of Banana Exporting Countries which is trying to raise the price of bananas.
B The European Community gives interest free loans to Dominica if its earnings from bananas fall.
C Dominica starts more manufacturing industries, making juices, jams and jellies from all the fruit it grows.

Fig. 13.6

Case Study of Zambia, Africa

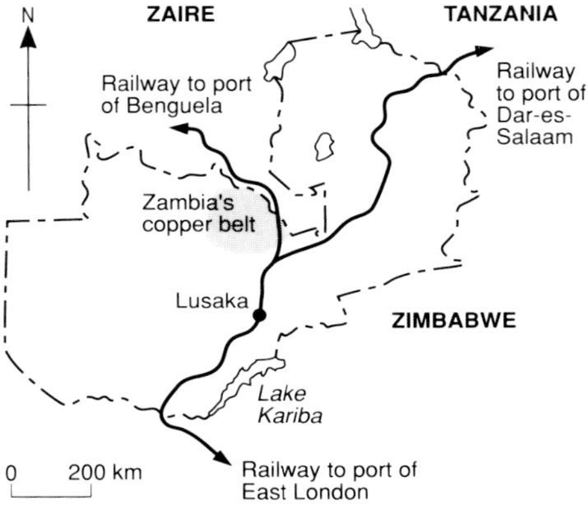

Fig. 13.7

Exports (1985)	Imports (1985)
Copper (89%)	Machinery
Zinc	Vehicles
Lead	Textiles
Cobalt	Iron and steel
Tobacco	Petroleum products

Fig. 13.9

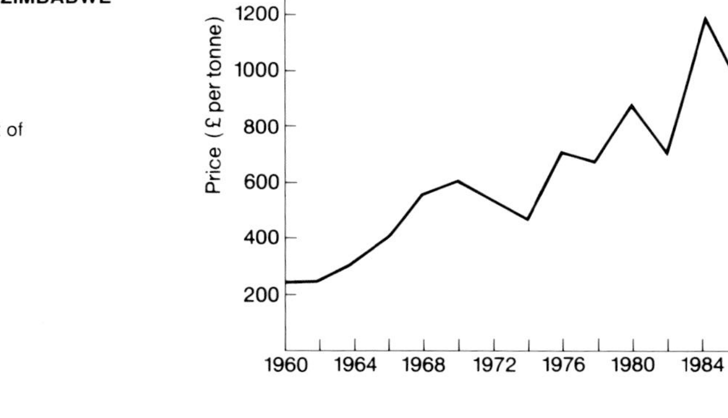

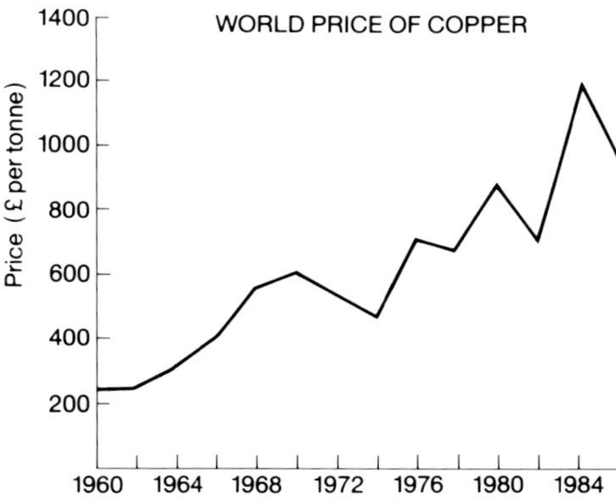

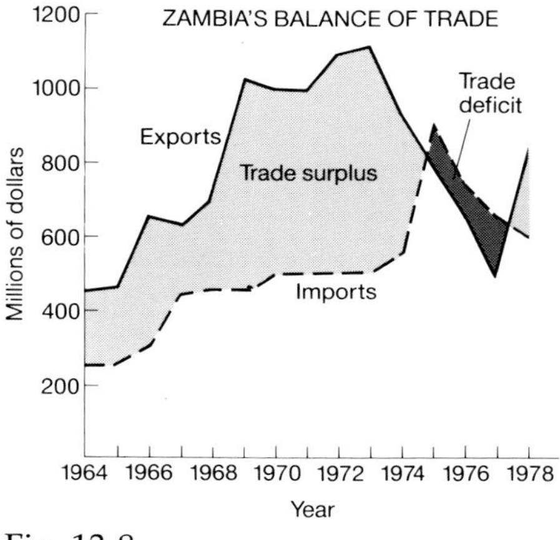

Fig. 13.8

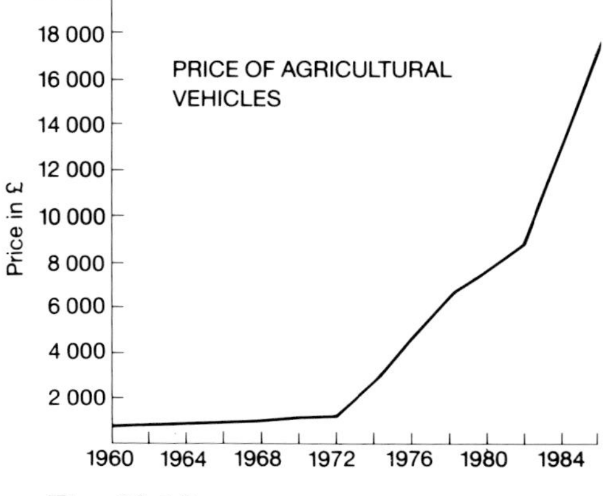

Fig. 13.10

Solutions to Zambia's Trade Problem

A Start manufacturing industries eg making copper wire, cigarettes.
B Zambia joins a group of Copper Exporting Countries to get stable high prices for copper.
C Interest free loans from the European Community when exports fall by 10 per cent or more.

Fig. 13.11.

A copper mine in Zambia

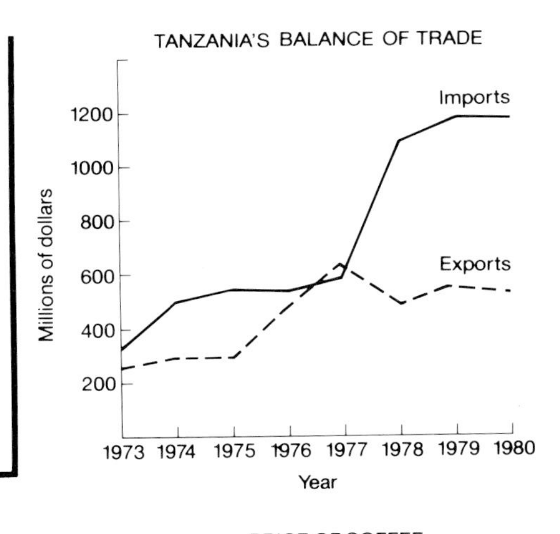

Case Study of Tanzania, Africa

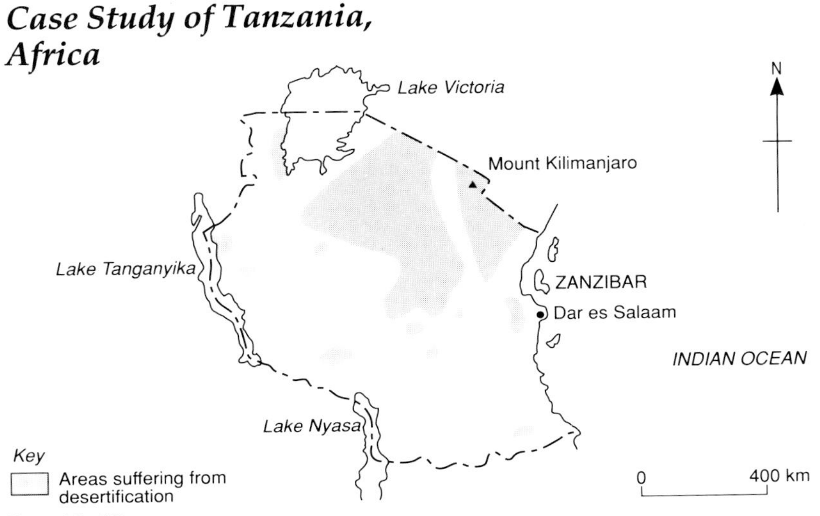

Fig. 13.12

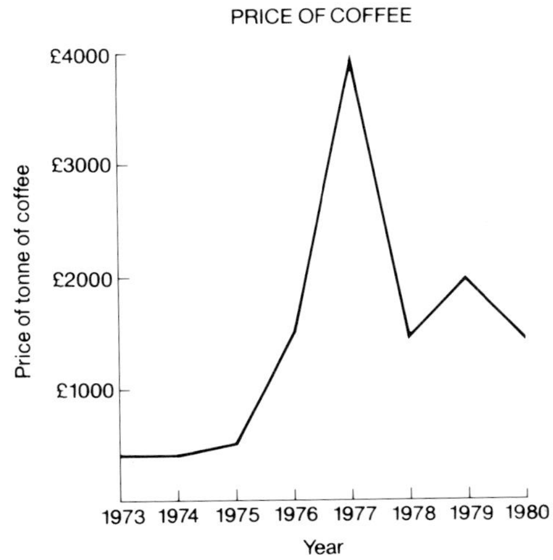

Fig. 13.16

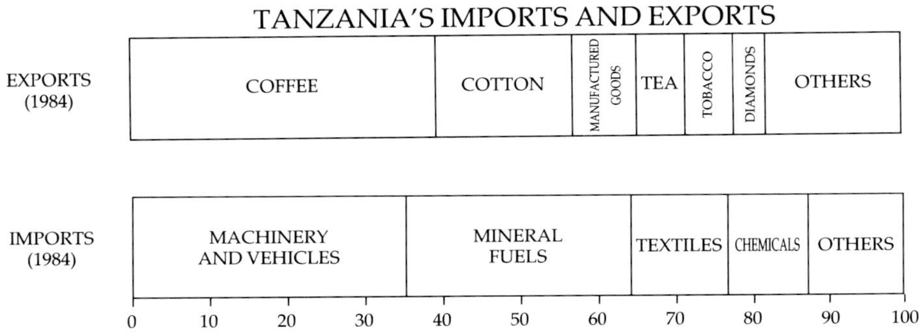

TANZANIA'S IMPORTS AND EXPORTS

EXPORTS (1984): COFFEE | COTTON | MANUFACTURED GOODS | TEA | TOBACCO | DIAMONDS | OTHERS

IMPORTS (1984): MACHINERY AND VEHICLES | MINERAL FUELS | TEXTILES | CHEMICALS | OTHERS

0 10 20 30 40 50 60 70 80 90 100

Fig. 13.13

Exports to UK (1988) £26 million (10% of Tanzania's exports)

Imports from UK (1988) £88 million (0.1% of UK's exports)

Fig. 13.14

Tanzania built an instant coffee factory, using local coffee beans. But to set up the factory, provide the equipment and train the labour force, it needed the help of a multinational company. Tanzania hopes to set up more factories in the future.

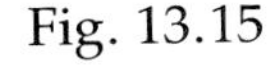

Fig. 13.15

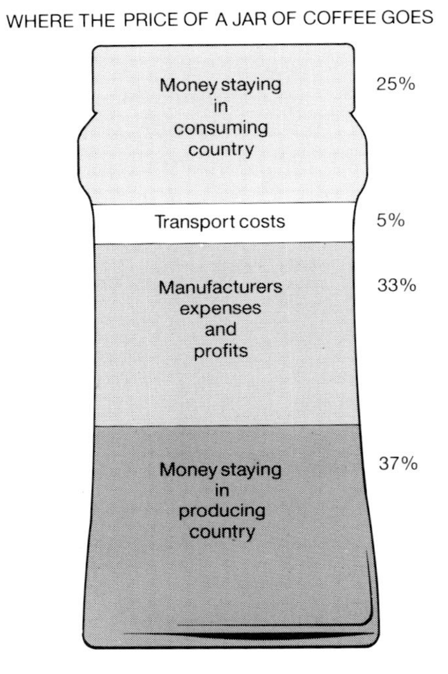

Fig. 13.17

Extension Text

13F Characteristics of International Trade

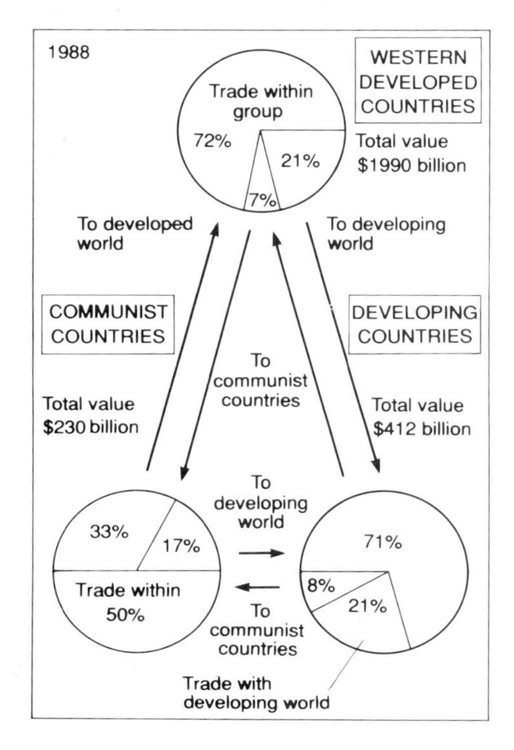

13G Dependence and Interdependence

All countries are **producers** of goods and **consumers** of goods. All countries export some of the goods they produce and import some of the goods they consume.

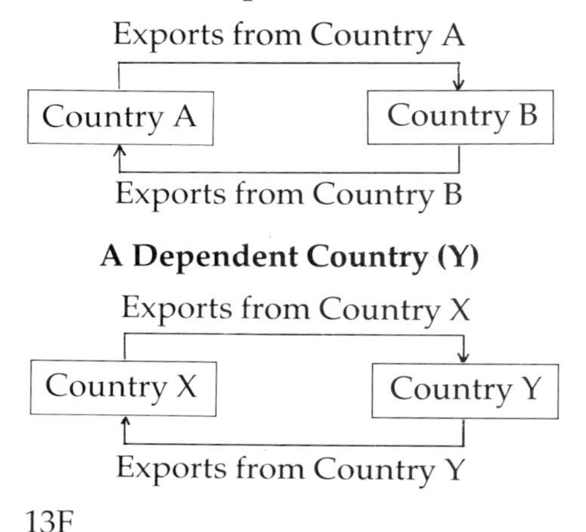

13F

If two countries trade a lot of goods with each other they are **interdependent** – they need each other equally. But if one country exports much more than it imports from another country it is **dependent** on that country.

Developing Countries are dependent on Developed Countries because they rely on a few Developed Countries to buy most of their exports. This makes it very difficult for them to be politically independent.

If a Developing Country loses an export market, it badly affects the whole country because it has so few exports and markets. Unemployment would rise sharply and there would be little money for any development. If a Developed Country loses an export market, it could badly affect regional employment and income eg the shipbuilding industry of the UK.

13H Multinational or Transnational Companies

Developing Countries may not just be dependent on another country for their exports. They may also be dependent on a foreign company to develop and export their resources.

Companies with branches in many countries are called **multinationals** or **transnationals**. Their annual sales can be greater than the gross national product of many Developing Countries. Multinationals bring both advantages and disadvantages when they set up in a Developing Country.

Advantages
Provide jobs
Improve the skills of the people
Provide the money and technology to develop the country's resources
Improve the services eg roads, ports
The country takes a share of the profits

Disadvantages
Jobs are often poorly paid
Many skilled workers are not local people
Farming is very mechanized and few workers are needed
Most of the profits go overseas
They may pull out at any time

Questions

Look at 13F, 13G and 13H.

E1 Describe the trade pattern of (a) the Western Developed Countries and (b) the Developing Countries.
E2 In what ways can a Developed Country affect employment in a Developing Country?
E3 Why is it difficult for a Developing Country to be politically independent?
E4 What is a multinational company?
E5 Explain how multinationals can benefit a Developing Country.
E6 In what ways can a Developing Country be dependent on a multinational company?

Questions

Case Study of Tanzania, Africa

Look at fig. 13.13.
C1 Compare the relative importance of primary and manufactured goods in Tanzania's imports and exports.

Look at fig. 13.16.
C2 Describe the relationship between Tanzania's balance of trade and the world price of coffee.

Look at fig. 13.14.
C3 Do you think Tanzania and the UK are interdependent? Give reasons for your answer.

Look at fig. 13.16.
C4 Do you think Tanzania's money from coffee stays the same from year to year? Give reasons for your answer and describe the consequences for Tanzania.

Look at figs. 13.15 and 13.17.
C5 What are the advantages and disadvantages of Tanzania setting up its own instant coffee factory?
C6 The UK puts tariffs on instant coffee, but not on raw coffee beans.
How will the opinion of people in the UK and Tanzania differ about the use of tariffs?

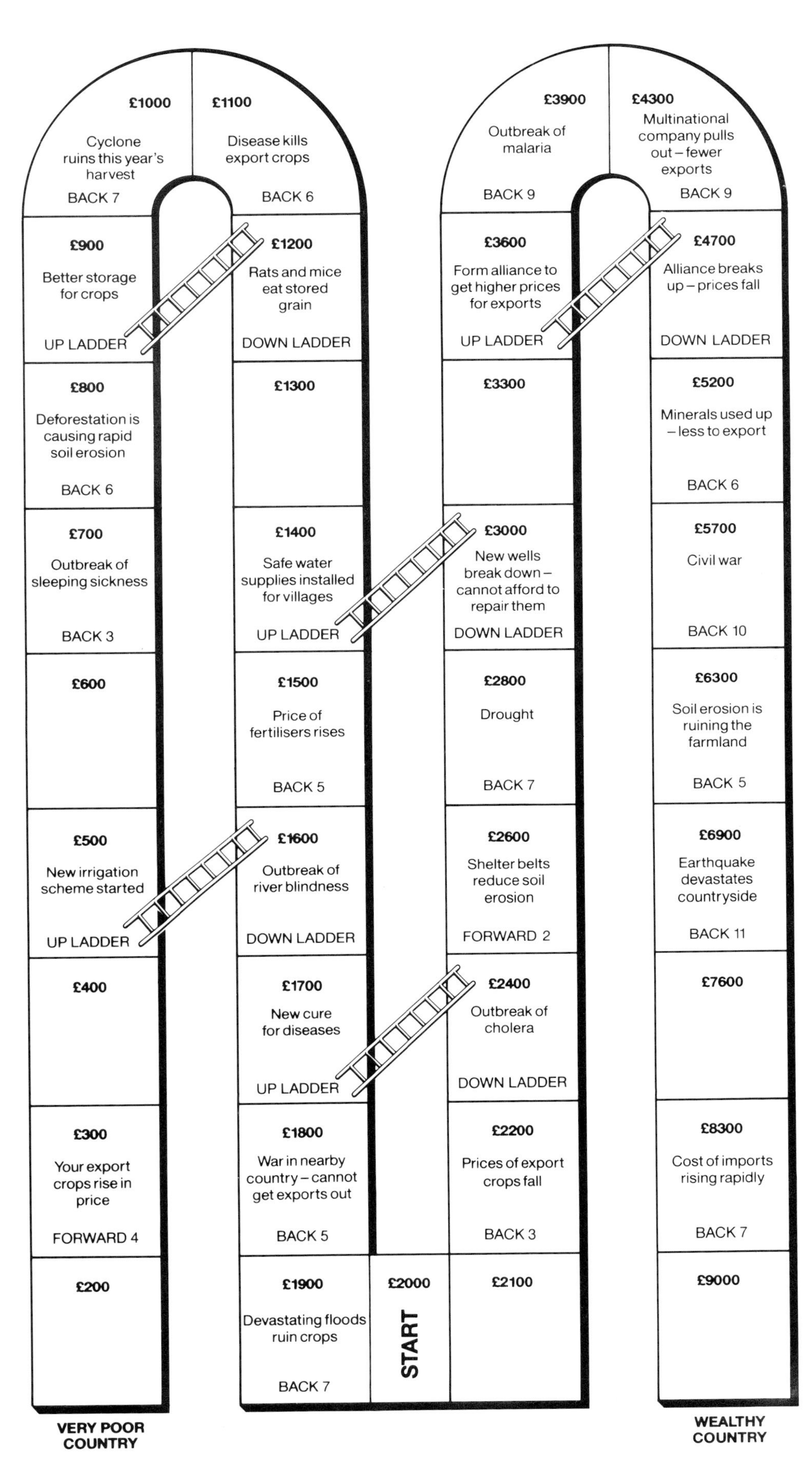

Core Groupwork

The Development Game

This game is about a Third World Country trying to develop.

TASK 1 You need a partner to play with and another couple to play against.

You need a dice and a counter for each couple.

TASK 2 Agree how many turns to have each.

One turn represents one year in the country.

TASK 3 Place the counter at the **start**. The start is where the people in the country have an average income of £2000 per year.

TASK 4 Each pair throws the dice in turn, moves and follows the instructions on the square it lands on.

TASK 5 While one partner throws the dice, the other partner writes down the following information:

Year	Average Income	Reason for Rise or Fall
1		
2		
3		
⋮		

TASK 6 When you have agreed to end the game, draw a line graph of your income over the years.

TASK 7 Write a report, explaining why your country's income has gone up and down.

UNIT 14 International Aid and Self-Help

Core Text

14A Aid and Self-Help

All countries try to raise the standard of living of their people (called **development**). But as the last units have shown, development brings problems. Sometimes the people can solve the problems themselves (called **self-help**). Sometimes they need help from other countries (called **international aid**).

14B Types of International Aid

Short-term aid is given when help is needed urgently.

Short-term problem	Short-term aid
Floods	Clothing
Famine	Food
Earthquakes	Shelter
Hurricanes	Medicine
War	

INTERNATIONAL AID SCHEME

Often needs experts from other countries to build and maintain

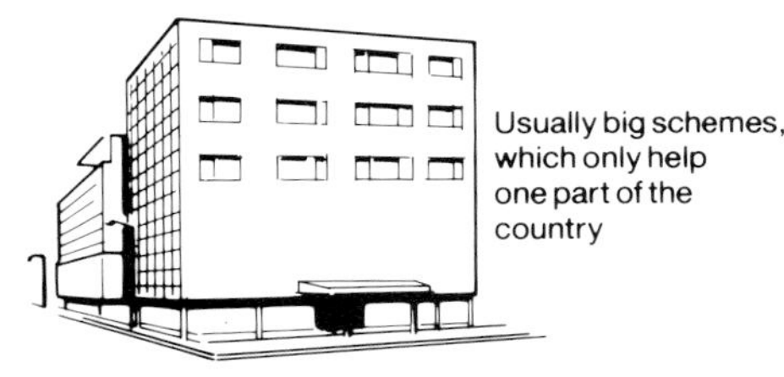

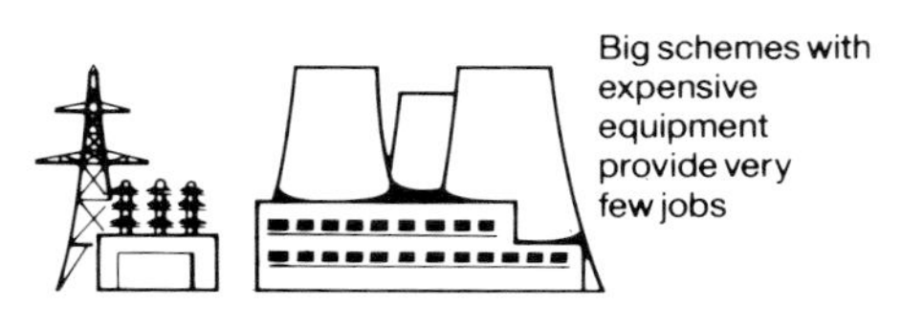

Long-term aid is given to help countries to develop. It may take months or years before the aid brings benefits.

Long-term problem	Long-term aid
Poor farming	Machines, power-stations
Bad health care	New hospitals
Poor transport	New roads
Many people not being able to read or write	New schools
Increasing populations	New houses

14C Self-Help Schemes

International aid brings many benefits, but it can bring problems as well. Instead, many countries are now starting up **self-help schemes.**

SELF-HELP SCHEME

Uses local materials and is made in workshops by local people

Training for local people in every village eg health workers, farming advisers

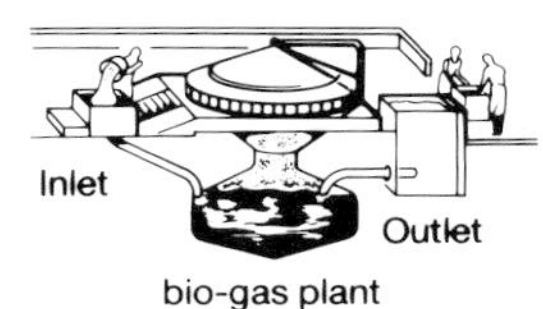

Small workshops in every village, using the skills of the local people, provide many jobs

14D Where International Aid comes from

Official Aid is when one country or group of countries (eg United Nations) gives aid to another country. The money comes out of taxes.

Voluntary Aid is the aid given by charities eg Band Aid, Oxfam, Save The Children Fund. They raise their money from people who make gifts and donations. Some charities also receive funds from government.

14E The United Nations – Official Aid

159 countries give money to the United Nations. The different departments or **agencies** of the UN then use this money for different purposes. The different agencies are listed below.

World Health Organisation (WHO)

United Nations Disaster Relief Organisation (UNDRO)

United Nations Educational Scientific and Cultural Organisation (UNESCO)

United Nations International Children's Emergency Fund (UNICEF)

Food and Agriculture Organisation (FAO)

14F Oxfam – Voluntary Aid

Between 1983-1988, for every £1 given Oxfam spent an average of 3p on administration, 4p on shop costs and 12p on fundraising costs

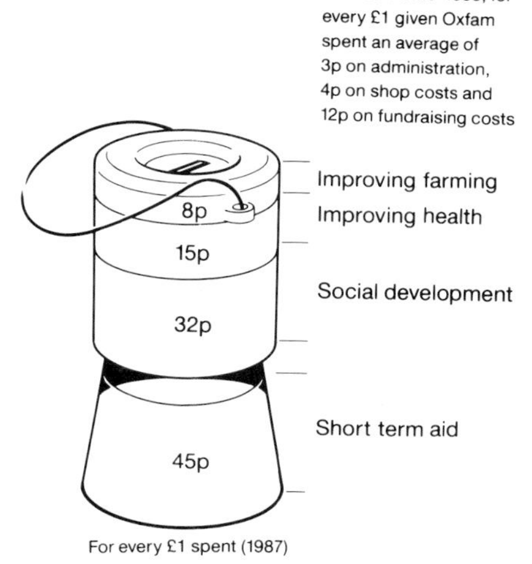

For every £1 spent (1987)

Core Questions

Look at 14B.
1 Give two reasons why a country might need short-term aid.
2 Name three examples of short-term aid.
3 Which of the following are examples of long-term aid:
(*a*) a new hospital,
(*b*) pumps to get rid of floodwater,
(*c*) new hotels for tourists,
(*d*) food and medicines?

Look at 14C.
4 Which schemes (international aid or self-help) are more likely to:
(*a*) be big, expensive schemes,
(*b*) employ more local people,
(*c*) use local materials,
(*d*) help all of the country?

The money needed to provide enough food, water, education, health and housing for everyone in the world has been estimated at £10 billion a year. It is a huge sum of money . . . about as much as the world spends on arms every two weeks.

5 What is the poster above trying to say?

Look at 14D.
6 Name two charities.
7 From where do charities get their money?
8 From where does a country get the money that it gives as official aid?

Look at 14E.
9 Which United Nations agency would help with these problems:
(*a*) homeless families after an earthquake,
(*b*) an outbreak of disease,
(*c*) not enough food being grown?

Look at 14F.
10 Does Oxfam spend more money on short-term or long-term aid?

Questions

Case Study of Yemen Arab Republic

Look at fig. 14.3.
F1 Describe some of the problems in Yemen.

Look at fig. 14.4.
F2 In what way has the amount of aid to Yemen changed?

'Why should we give aid to Yemen? We don't get anything from them'

Look at fig. 14.6.
F3 The statement above is exaggerated. Explain how it is exaggerated.

Look at fig. 14.2.
F4 After the earthquake in Yemen, charities gave short-term and long-term aid. Do you think they should have spent all their money on short-term aid? Give a reason for your answer.

Look at figs. 14.1 and 14.3.
F5 Choose from (a) a new hospital (b) more wells and (c) new roads. Which of these schemes would most help the people of Yemen? Give reasons for your answer.

Look at fig. 14.5.
F6 Which scheme – international aid or self-help – will do more to help the people of Yemen? Give reasons for your answer.

Questions

Case Study of Pakistan

Look at figs. 14.7 and 14.10.
G1 Describe some of the problems faced by Pakistan.
G2 For which of the problems in Fig. 14.11 is short-term aid most needed? Give reasons for your answer.

Poor farming methods causing food shortages in Pakistan

'Spend all the money on short-term aid for the people'.
'No, use some of the money for long-term aid.'

Look at the newspaper headline and statements above.
G3 Explain both of the points of view above.
G4 Give one advantage and one disadvantage of the international aid scheme described in Fig. 14.12.

Look at fig. 14.9.
G5 What are the advantages and disadvantages to Pakistan of official aid over voluntary aid?
G6 If there was only enough money for one of the schemes described in Fig. 14.8, which do you think should be chosen? Give reasons for your answer.
G7 What is the cartoon below trying to show about the use made of offical aid to Pakistan?

Resources

Case Study of Yemen Arab Republic

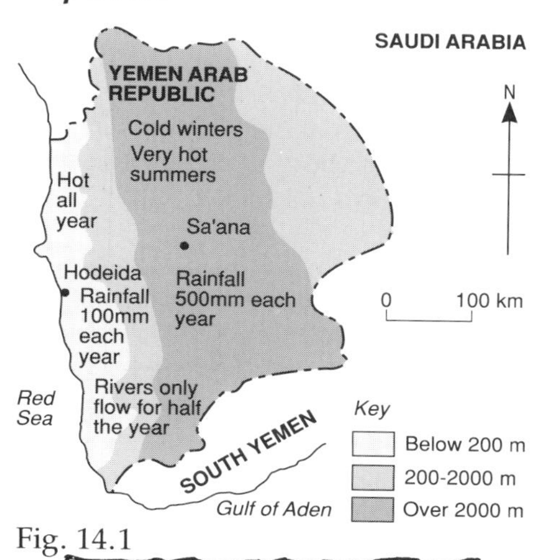

Fig. 14.1

Earthquake in Yemen kills 3000, leaves 400 000 homeless. Roads, houses and farmland in ruins.

Fig. 14.2

The Yemen Arab Republic is a very poor, desert country. Most of the people don't get enough to eat. There are few doctors and only 1 in 5 people have clean water to drink. Polluted water kills many children and snail fever is very common. There are few schools and only 1 in 5 adults can read and write.

Fig. 14.3

Aid from the UK to Yemen	
1985	£3 100 000
1986	£3 800 000
1987	£5 200 000

Fig. 14.4

Developments in Yemen
International aid scheme: a new irrigation scheme in the Yemen coastal plain grows cotton for export.
Self-help scheme: in villages over the Yemen Highlands, people are being trained to advise farmers on better crops, animals and ways to farm.

Fig. 14.5

Yemen buys £43 million of goods from UK each year. Yemen provides UK with cotton and coffee.

Fig. 14.6

Case Study of Pakistan

Fig. 14.7

A Self-Help Scheme
Health centres in 30 villages were set up and one person in each village was trained to treat common diseases and give family planning advice. Cost £20,000.

An International Aid Scheme
Medical and surgical equipment supplied to one city hospital. Cost £20 000.

Fig. 14.8

AID TO PAKISTAN

Official Aid (90%)
1 Loans to be repaid with interest
2 'Tied Aid' – money has to be spent on goods from the country giving the money
3 Grants for big projects which do not help the poorest

Voluntary Aid (10%)
1 Only enough for small projects
2 Used to help people in most need
3 Depends on funds being raised

Fig. 14.9

Facts about Pakistan

Calories eaten per day	2159
Calories needed to be healthy	2500
Exports	£2300 million
Imports	£3300 million
Birth rate	42%
Death rate	10%

Fig. 14.10

Some Problems in Pakistan
1 Only 17% of children go to secondary school.
2 Three million refugees from Afganistan just settled in Pakistan.
3 Lack of roads in northern Pakistan.
4 Most families do not have a clean water supply.

Fig. 14.11

International aid has provided farmers in Pakistan with high yielding seeds of wheat and rice. These provide a lot more food than other seeds but they need a lot more fertiliser. Only the richer farmers can afford the fertiliser.

Fig. 14.12

Case Study of Sri Lanka

The Mahaweli River Project
1 Cost £1650 million – some money from increased taxes; most came from 'tied aid'.
2 Four dams and power stations were built.
3 The water is used to irrigate thousands of hectares of land.
4 The electricity saves using firewood and imported oil.
5 One million people had to be displaced from the area.
6 The reservoirs are stocked with fish.

Fig. 14.13

Developments in Sri Lanka
1 Self-help village projects, using appropriate technology to improve farming.
2 A new surgicial and maternity hospital in the city of Colombo.
3 Training of primary health care workers in every village.
4 A new hotel complex on the coast.

Fig. 14.14

Tea makes up 30% of exports
Birth rate 24‰
Death rate 6‰
GNP/head $400
People in agriculture 52%
Rapid deforestation

Fig. 14.15

Extension Text

14G Forms of Aid

International aid can be provided in many different forms, as the diagram below shows.

TECHNICAL AID
Advisers, consultants, specialists

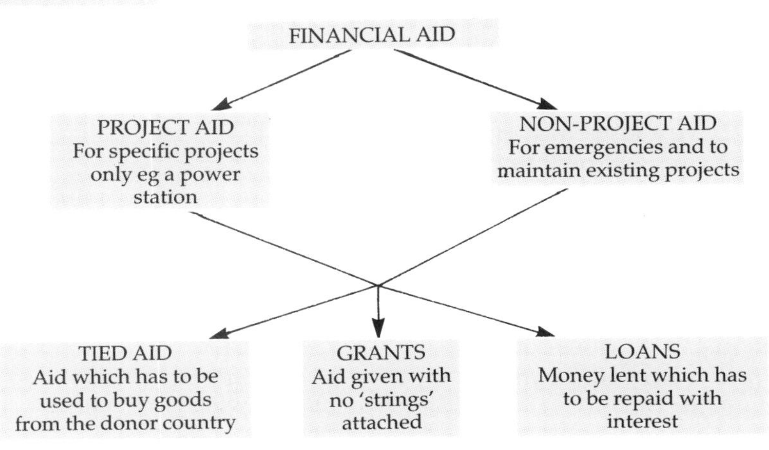

14H Sources of Aid

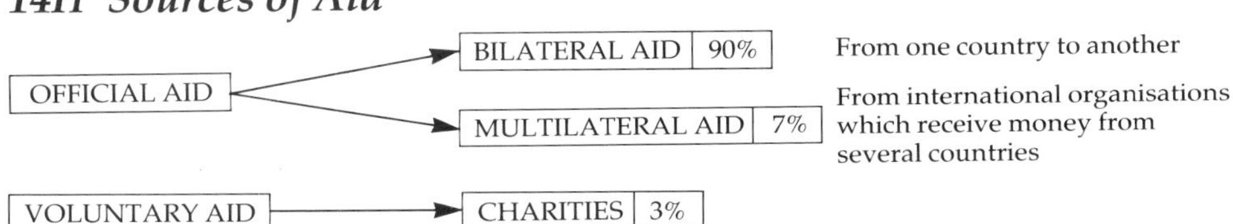

	Advantages	Disadvantages
Bilateral Aid	1 Quick and direct 2 Has large funds for major schemes	1 Mostly 'tied aid' 2 Receiving country is tied politically to donor country 3 Often used for large prestige projects using advanced technology
Multilateral Aid	1 Not 'tied' 2 No political strings	1 Mostly loans, which have to be repaid with interest 2 Takes longer for the aid to be given
Voluntary Aid	1 No political strings 2 Encourage self-help schemes, using appropriate technology	1 Very limited funds 2 Funds are less reliable, as they depend on gifts and donations

14J Aid and Technology

Advanced Technology uses specialist equipment and machinery; requires skills that most people in the Developing World do not have.

Intermediate or **Appropriate Technology** uses simpler equipment that local people are skilled enough to use.

Primitive Technology uses very simple methods; easy for people to work, but not very effective.

Questions

Look at 14G, 14H and 14J.

E1 What is financial aid used for?

E2 What form of aid is the cartoon above trying to show?

E3 What is the difference between bilateral and multilateral aid?

E4 Which source of aid most uses advanced technology schemes?

E5 Explain why intermediate technology is often called appropriate technology.

Questions

Look at fig. 14.15.

C1 Describe some of the problems faced by Sri Lanka.

Look at fig. 14.13.

C2 People in Sri Lanka have different attitudes towards the Mahaweli River Project. Describe some of the different points of view.

Look at fig. 14.14.

C3 If Sri Lanka had to choose between bilateral aid for projects 2 and 4 or voluntary aid for projects 1 and 3, which should she choose? Give detailed reasons.

Rapid deforestation in southern Sri Lanka is causing soil erosion and flooding. People are suffering from starvation and malnutrition.

C4 International help for the problem above could be spent on short-term aid, long-term aid or both. Which type(s) of aid should be given? Give detailed reasons why.

UNIT 15 The Influence of Countries in the World

Core Text

15A International Influence

If a country is going to develop, it needs help from other countries. It needs them to trade with, to provide them with aid or even to provide them with defence.

Countries which provide a lot of trade, aid and defence to others have a lot of **international influence.** They can influence the way other countries are run.

It is difficult to work out how much influence a country has. Some of the measures used are shown in 15B.

15B The Superpowers

The countries with the most international influence are called **superpowers.** The two biggest superpowers are the USA and the USSR. They can influence countries in many ways:

A By Trade – countries who sell most of their goods to the superpowers depend on them.

B By Aid – superpowers give most aid to countries they support.

C By Military Support – they can provide arms and advice. This can make the government of a country stronger. They can also make governments weaker by giving arms to opposition forces.

D By invasion – they directly change the government of a country.

15C Measuring International Influence

1 The Size of the Country
Large countries can have a lot of influence. They control large areas of land and all the resources on it and underneath it. They do not need to depend as much on other countries. But not all large countries have a lot of influence. Some have very few people and very few resources.

TOP 5 COUNTRIES (SIZE)
1 USSR
2 Canada
3 China
4 USA
5 Brazil

2 The Population of the Country
Countries with a large population can have a lot of influence. A large population can produce a lot of wealth and the country can also have a large number of armed forces. But not all countries with a large population have a lot of influence. In some countries the people are not healthy enough or educated enough to develop the country fully.

TOP 5 COUNTRIES (POPULATION)
1 China
2 India
3 USSR
4 USA
5 Indonesia

3 The Resources of a Country
Countries with a lot of resources can have a lot of influence. The resources can be used to make the country wealthy. But some countries are too poor to use all the resources they have. And some resources are more important than others. Coal, iron ore and oil are valuable because so many industries need them.

TOP 5 COUNTRIES FOR CRUDE OIL
1 USSR
2 USA
3 Saudi Arabia
4 Mexico
5 UK

4 The Military Strength of a Country
Countries with a lot of arms and armed forces have a lot of influence. They can provide arms or military advice to other countries. They can even attack other countries.

TOP 5 COUNTRIES (MILITARY STRENGTH)
1 USSR
2 USA
3 UK
4 Germany
5 France

5 The Wealth of a Country
Countries with a large amount of wealth can have a lot of influence. They can afford to buy goods from other countries and provide a lot of aid. They are also rich enough to build up huge military strength.

TOP 5 COUNTRIES (WEALTH)
1 USA
2 Japan
3 USSR
4 Germany
5 France

Core Questions

Look at 15C.

1 Name the five ways of measuring a country's influence.

2 Explain why some large countries have little influence.

3 Explain why some countries with a large population have little influence.

Look at 15B.

4 Which are the two biggest superpowers in the world?

5 Name three ways in which a superpower can influence another country.

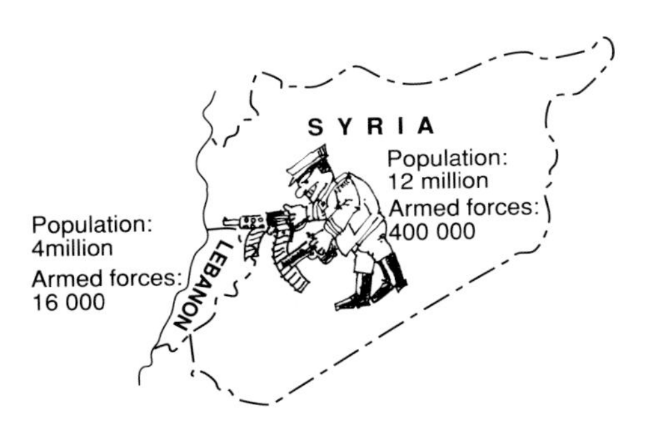

6 Explain what the cartoon above is trying to show.

Questions

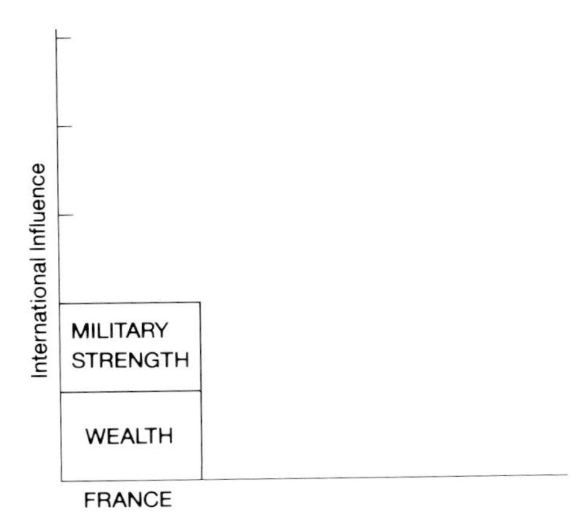

Look at the five tables of Top 5 Countries in 15C.

F1 Draw a bar chart for every country mentioned in the tables. The bars should be drawn 1 centimetre high for every table that country is in.

The bar chart for France is shown above. It is drawn 2 cm high because France appears in two tables.

F2 Describe what your bar charts show.

Questions

Look at the five tables of Top 5 Countries in 15C.

G1 Draw a circle for each country mentioned in the table. The circles should have a radius of 1 centimetre for every table that country is in.

G2 Describe what your circles show.

Questions

Look at the five tables of Top 5 Countries in 15C.

C1 Devise a points system to show the overall international influence of each country mentioned in the tables. Use a suitable diagram or diagrams to show your results.

C2 Describe what your results show.

Core Investigation

The next two pages give information on the international influence of the USA and the USSR. Decide what you are going to investigate about the USA and/or the USSR.

Think of an exact aim or aims. Plan out your investigation in detail – decide what you are going to find out firstly, secondly and so on.

Your report should be in 4 parts:

1 Aim
2 Method
3 Analysis
4 Conclusion

Use other books and atlases to get more information and maps. Look at page 16 to find out how to write up your report.

Resources

The United States of America

Historical Background

The USA was once a colony of Britain but broke away from Britain in 1776. At that time there were 13 states. Now there are 50 states, stretching from the Arctic Circle to the Tropics. Millions of people from all over the world have emigrated to the USA.

The USA is a **capitalist** country. This means that the land and factories are owned by different people and companies. They compete with each other to exploit the resources and help the country to develop. As a result, many people are very wealthy, but some are very poor.

The USA is also a **democracy.** This means that all adults have the right to vote for different political parties. The political party that wins the most votes runs or governs the country for the next four years.

The American system has advantages and disadvantages and it is changing all the time. In recent years, the government has had a bigger influence on how the country should develop.

10% of the USA is too cold to farm

20% of the USA is too mountainous to farm

Population (1989)
Number of people = 248 million

People who can read and write = 99%
Number of people per doctor = 520
People under 15 years = 23%
People 15 – 60 years = 61%
People over 60 years = 16%
Birth rate = 16‰
Death rate = 9‰

Resources (1985)
Minerals:
1st in the world for copper
1st for natural gas
2nd for coal
2nd for lead
2nd for crude oil.

Crops:
1st in the world for maize
2nd for oats
2nd for wheat
Animals:
2nd in the world for cattle.

The USA has immense resources but still has to import some commodities eg crude oil, paper and cars.

Imports per person = £800
Exports per person = £535

Wealth (1985)
Average income per person = £8100 per year
Aid given = £10,400 million per year.

The USA has the wealth and technology to improve its land. In 1984, over 18 million hectares of farmland were under irrigation.

Military Strength (1989)
Numbers in the Army = 770 000
the Navy = 587 000
the Air Force = 579 000

TANKS	16150
SUBMARINES	127
COMBAT AIRCRAFT	5475
NUCLEAR WARHEADS	9680

Size
Area = 9 200 000 sq km

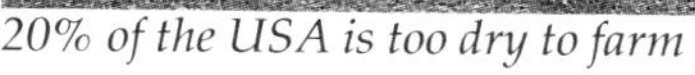

20% of the USA is too dry to farm

The Union of Soviet Socialist Republics (the Soviet Union)

Historical Background
The USSR came into being in 1917, when the King (Tsar) of Russia was overthrown in the Bolshevik Revolution. Russia is now just one of 15 republics which make up the Soviet Union. Over 100 different nationalities live in the Soviet Union.

The USSR began as a **communist** country. This means that the State owned all the land and factories and decided how the country should develop. The wealth is shared out more equally, so there are not many very rich people in the USSR nor are there many very poor people.

The USSR began as a **One Party State.** The Communist Party was the only political party. The people could only vote between different candidates in the Communist Party.

The Soviet system has both advantages and disadvantages and is changing all the time. In recent years, the people have been given a lot more freedom and choice.

50% of the USSR is too cold to farm

Population (1990)
Number of people = 289 million

People who can read and write = 99%
Number of people per doctor = 280
People under 15 years = 24%
People 15-60 years = 66%
People over 60 years = 10%
Birth rate = 18‰
Death rate = 9‰

10% of the USSR is too mountainous to farm

Resources (1985)
Minerals:
1st in the world for coal
1st for crude oil
1st for iron ore
1st for lead
2nd for copper
2nd for natural gas.

USSR has immense wealth, but she still has to import some important commodities, especially ships, railways, minerals.

Crops:
1st in the world for barley
1st for oats
1st for wheat
Animals:
1st in the world for sheep
2nd for fish.

Imports per person = £170
Exports per person = £194

Wealth (1985)
Average income per person = £1500 per year
Aid given = £1000 million per year.

The USSR has the wealth and technology to improve its land. In 1986 over 20 million hectares of farmland were under irrigation.

Military Strength (1989)
Numbers in the Army = 2 million
the Navy = 460 000
the Air Force = 440 000

TANKS	61700
SUBMARINES	323
COMBAT AIRCRAFT	7250
NUCLEAR SUBMARINES	10996

Size
Area = 22 400 000 sq km

15% of the USSR is too dry to farm

UNIT 16 Conflicts and Flashpoints

Core Text

16A How Conflicts Become Flashpoints

In many countries all over the world there are **conflicts.** Conflicts happen when people and countries have different views. Some conflicts are more serious because other countries become involved. These are called **flashpoints.**

There are several reasons why conflicts become flashpoints:

16B World Flashpoints

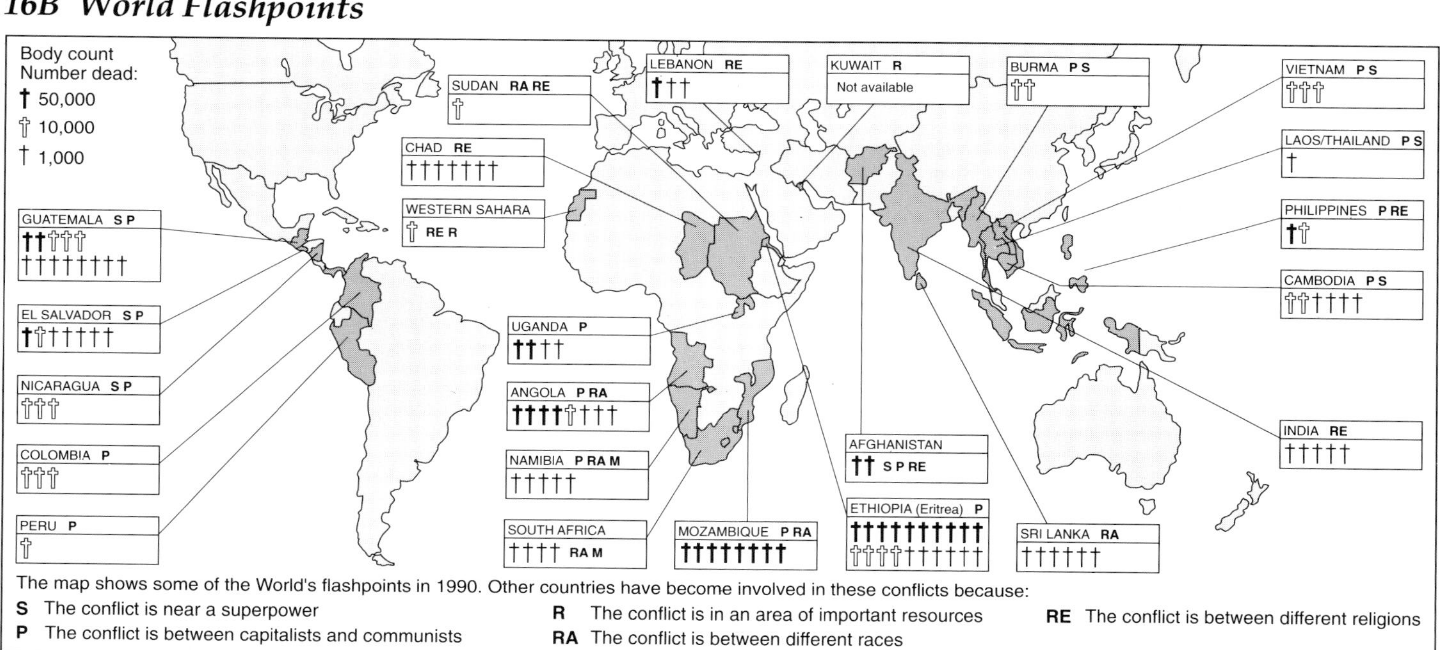

The map shows some of the World's flashpoints in 1990. Other countries have become involved in these conflicts because:

S The conflict is near a superpower
P The conflict is between capitalists and communists
R The conflict is in an area of important resources
RA The conflict is between different races
RE The conflict is between different religions

Core Questions

Look at 16A.
1 Why do conflicts happen?
2 How do conflicts become flashpoints?
3 The cartoon shows four ways in which other countries get involved in conflicts. What are the four ways?
4 The cartoon gives four reasons why other countries get involved in conflicts What are these reasons?

Look at the cartoon in 16A and the comments above.
5 Why does the President in each country not like
(*a*) sending in a peace-keeping force?
(*b*) supplying a lot of arms?
(*c*) stopping trade?
(*d*) sending money?

Look at 16B.
6 Name one area of the World which is a flashpoint partly because
(*a*) it is near a superpower,
(*b*) it has important resources,
(*c*) the conflict is between different religions,
(*d*) the conflict is between different races,
(*e*) the conflict is between capitalists and communists.

Questions

Case Study of Nicaragua, Central America

Look at fig. 16.1.
F1 Which countries in Central America have changed from capitalist to communist since 1958?
F2 Do you think the USA is pleased with the changes in government since 1958? Give a reason for your answer.

Look at figs. 16.2 and 16.8.
F3 Which two of the reasons in fig. 16.8 best explain why the USA was involved in Nicaragua?
F4 Which of the reasons in fig. 16.8 best explains why the USSR was involved in Nicaragua?

Look at fig. 16.3.

F5 The USA wanted to change the government of Nicaragua in the 1980s. Which of its methods in fig. 16.3 do you think did the most harm to the Nicaraguan government? Give reasons for your answer.

'If you supply arms to one group in Nicaragua, we will supply more arms to the other' (American General)
'If you supply arms to one group in Nicaragua we will supply more arms to the other' (Russian General)

F6 Read the statements above. What might happen next in Nicaragua? Give a reason for your answer.

Look at fig. 16.5.
F7 The USA and USSR turned the conflict in Nicaragua into a flashpoint. Do you think that they helped the people in Nicaragua? Give reasons for your answer.

Look at fig. 16.9.
F8 Which of the solutions would best stop the conflict in Nicaragua? Give reasons for your answer.

Questions

Case Study of Angola, Southern Africa

Look at fig. 16.10.
G1 Describe the changes in government in southern Africa since 1958.
G2 Do you think that South Africa is pleased with the changes in government since 1958? Give reasons for your answer.

Look at figs. 16.11 and 16.16.
G3 Which of the reasons in fig. 16.16 best explains why the USA became involved in Angola in 1975? Give reasons for your answer.
G4 Which of the reasons in fig. 16.16 best explains why the USSR became involved in Angola in 1975? Give reasons for your answer.

'South Africa and Cuba became involved in Angola for the same reasons'
G5 Give arguments for and against the point of view above.

Look at fig. 16.12.
G6 What are the advantages and disadvantages of the USA stopping trade with Angola?
G7 To change the government of Angola, South Africa could send in troops or supply military aid to the rebel forces. Explain the arguments for both these methods.

G8 What is the cartoon above trying to show?

Look at fig. 16.13.
G9 Which solution would best bring a lasting peace to Angola?

Resources

Case Study of Nicaragua, Central America

Fig. 16.1

The Conflict in Nicaragua
1970s Country ruled by a capitalist government under President Samoza
1979 President overthrown and new communist government takes over (called 'the Sandanistas')
1982 Rebel capitalist forces (called 'the Contras') try to overthrow the government USA gives help to Contras. USSR and Cuba give help to the government.
1990 Communist government voted out, after an election.

Fig. 16.2

The USA and Nicaragua
1 It placed mines in Nicaragua ports – this made it difficult for the country to import and export goods. The government could not improve the peoples' standard of living and it lost support
2 It stopped buying and selling goods from Nicaragua. This is known as an **embargo** – the government had little money because it could not sell as many goods
3 It gave arms to the Contra rebels – the rebels could now fight the government better

Fig. 16.3

The USSR and Nicaragua
1 It supplied arms to the government
2 It supplied much needed imports eg oil.

Fig. 16.4

Effects of the Nicaraguan Flashpoint
1977-79 50 000 killed when capitalist government overthrown
1982 USA and USSR became involved
1982-90 30 000 killed in fighting between the government and rebels and a quarter of a million people made homeless

Fig. 16.5

Conflict in El Salvador
This is between the capitalist government and communist rebels. 65 000 have been killed since 1979.

Fig. 16.6

Conflict in Guatemala
This is between the capitalist government and communist rebels. 138 000 have been killed since 1966 and 1 quarter of a million people have fled to Mexico.

Fig. 16.7

Why Other Countries are involved in Central America
1 Because Central America is nearby
2 To support the communists or capitalists
3 Because the countries have valuable resources

Fig. 16.8

Solving The Conflict in Nicaragua
1 Send in a United Nations peace-keeping force – they may stop the fighting while they are there
2 Hold an election – let the people vote for the government they want
3 Send in more arms until one side wins
4 Other countries stop providing arms to the two sides

Fig. 16.9

Case Study of Angola, Southern Africa

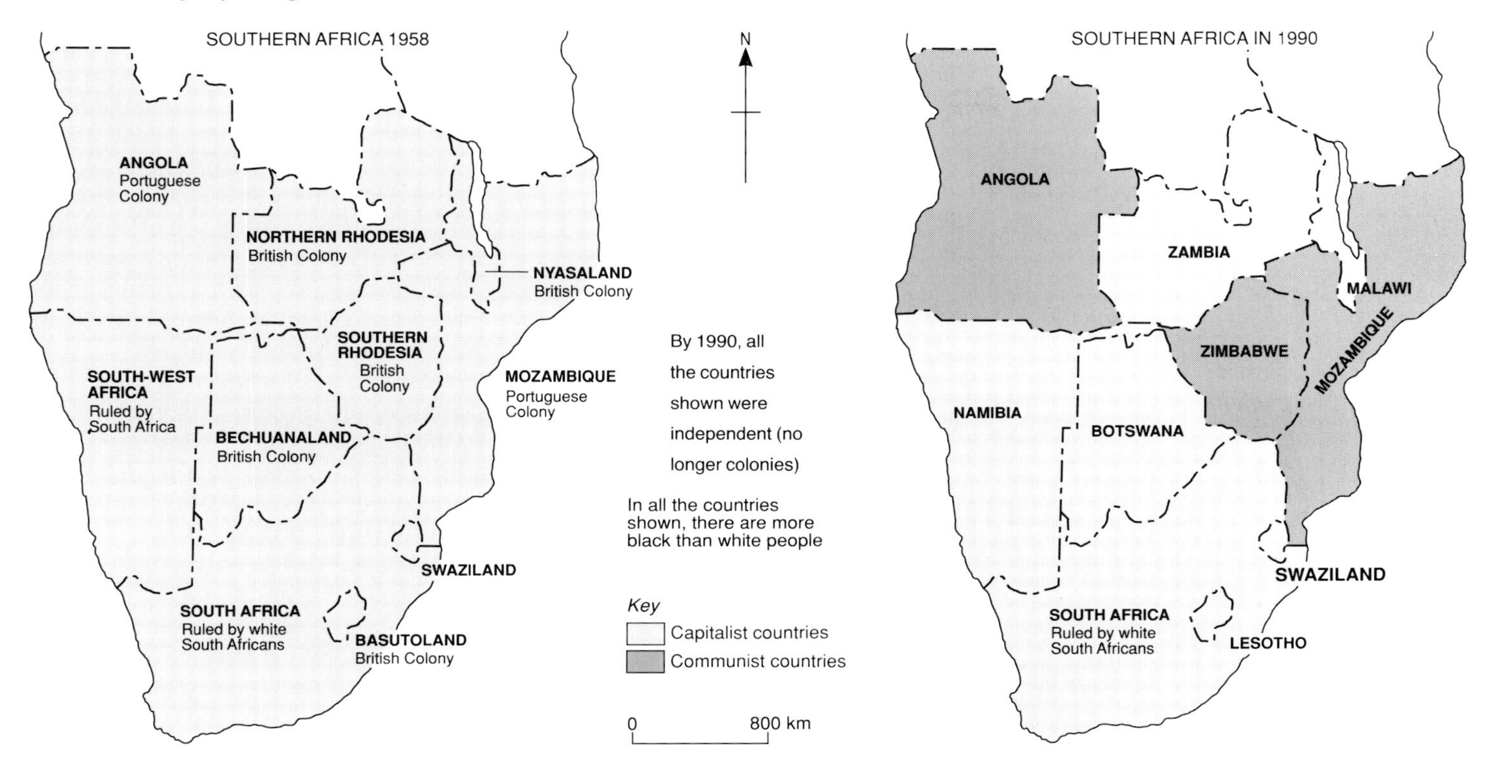

Fig. 16.10

The Conflict in Angola
Before 1975 It was a colony of Portugal
1975 Angola became independent. Start of Civil War between capitalist groups (called UNITA) and communist groups (MPLA). USA and South Africa give help to UNITA. USSR and Cuba give help to the MPLA
1976 Communist group takes over the government but fighting continues

Fig. 16.11

How other Countries are involved in Angola
1 Supplying military aid. This has to be paid out of the country's taxes and increases the amount of fighting
2 Sending in troops which costs lives and increases the amount of fighting
3 Stopping trade – Angolan government loses support because it cannot afford to improve the peoples's standard of living. It also reduces profits of companies in the country which stops trading and increases unemployment

Fig. 16.12

Solving the Conflict in Angola
1 Hold elections – the people vote for the government they want
2 Increase arms and troops until one side wins
3 Send in a United Nations peace-keeping force – they may stop the fighting while they are there
4 Other countries stop providing arms to the two sides

Fig. 16.13

Conflict in Namibia (South-West Africa)
In 1959 there was a civil war between the government, run by white South Africa and black Namibian rebels (called SWAPO)

In 1989 South Africa left and the people voted for a communist government

Fig. 16.14

Conflict in Mozambique
In 1975 a black communist government came to power. South Africa started to make raids into parts of Mozambique. USSR sent military aid to Mozambique.

Fig. 16.15

Why Other Countries are involved in Southern Africa
1 The countries are neighbours
2 There are minerals that other countries need
3 To support or oppose a political system
4 To support or oppose a racial group

Fig. 16.16

Effects of the Angola Flashpoint
In 1975 100 000 killed
Since 1975 At least 100 000 killed

Fig. 16.17

Case Study of Vietnam, South-East Asia

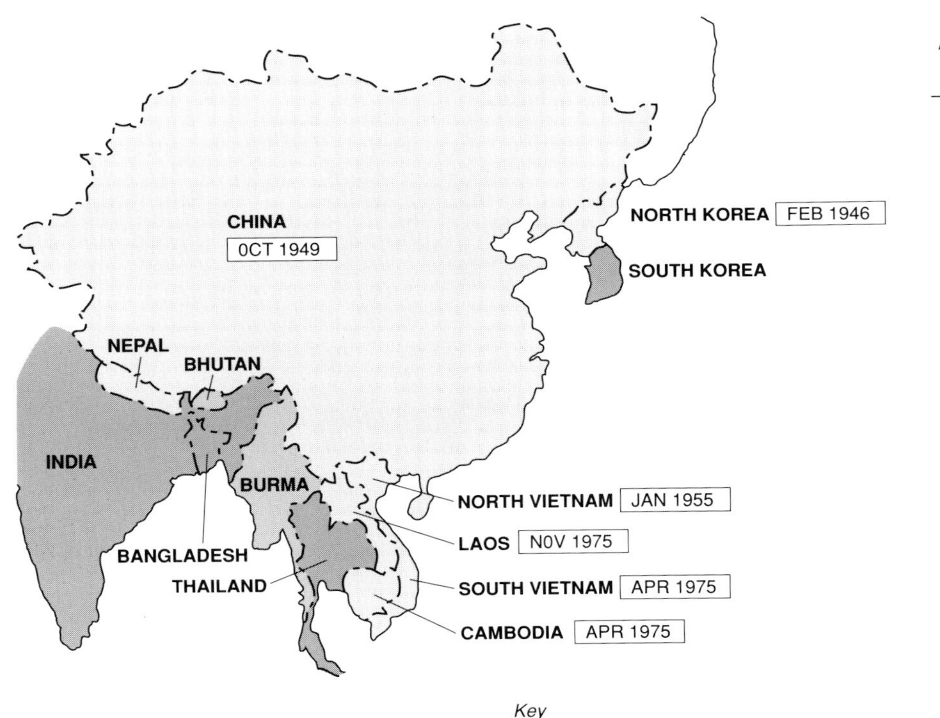

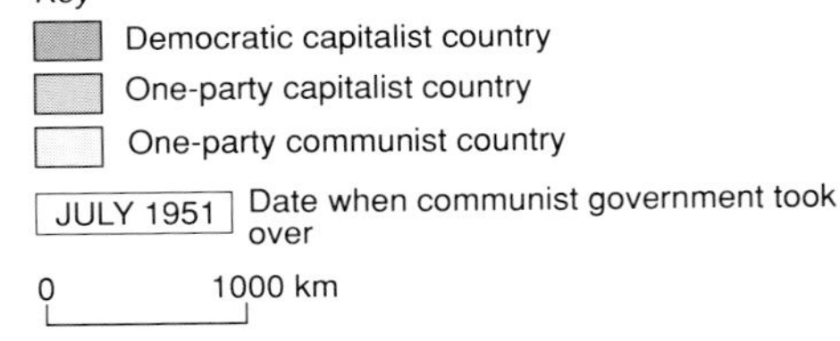

Fig. 16.18

The Conflict in Vietnam
1954 French rule of Vietnam ends
1955 Vietnam split into communist North Vietnam and capitalist South Vietnam
1961 War breaks out between North and South Vietnam. USA helps South Vietnam. USSR and China help North Vietnam
1970 Fighting spreads to Cambodia
1975 South Vietnam surrenders North and South Vietnam become one communist country
1978 Vietnam invades Cambodia (Kampuchea). USSR helps Vietnam. China opposes Vietnam
1989 Vietnam pulls out of Cambodia

Fig. 16.19

Possible United Nations Solutions to the Vietnam War
1 Send in UN peacekeeping force
2 Impose arms embargo on both countries
3 Impose trade embargo on both countries

Fig. 16.20

Effects of the Conflicts in Vietnam
1961-75 1 800 000 killed (mostly civilians)
Since 1979 30 000 killed

Fig. 16.21

VIETNAM – AFTER THE WAR
'We are now free of that inordinate fear of communism which once led us to embrace any dictator who joined our fear. This approach has failed – with Vietnam the best example of its intellectual and moral poverty.'
President Carter, 1977

Fig. 16.22

This speech by President Carter sums up the major change in American foreign policy which the Vietnam War has brought about. From 1947 the USA's foreign policy was dominated by the fear of communism's spreading throughout the world. This has led America into forming a world-wide system of alliances and intervening throughout the world to protect the 'free world' from communism. It also led her to defending the governments of unpopular and corrupt dictatorships. This policy came to grief in South-East Asia and Carter's speech is official recognition of America's changing role as 'world policeman'. Undoubtedly the ending of the war in Vietnam has made it easier for the USA to establish closer relations with the USSR and above all China.
Contemporary World, 1979

Fig. 16.23

The Conflict in Cambodia
1970 Cambodia became the Khmer Republic. Fighting between right-wing forces of the Khmer Republic and left-wing Khmer Rough forces
1975 Khmer Rouge forms communist government under Pol Pot
1975-1979 Over 3 million Cambodians killed by Khmer Rouge
1979 Vietnam overthrows the Khmer Rouge and takes over the government
1989 Vietnamese forces leave Khmer Rouge guerillas still fighting

Fig. 16.24

How the USA was involved in Vietnam
1959 It sent military advisers to South Vietnam
By 1961 15 000 US troops were in Vietnam
By 1967 500 000 US troops were in Vietnam
By 1975 57 000 US troops had been killed
1961-1975 The war cost USA £30 000 million each year

Fig. 16.25

Extension Text

16C Why Conflicts Become Flashpoints

Local conflicts escalate into flashpoints when other countries especially the superpowers, become involved. They become involved for one or more of the following reasons:

1 *The Strategic Importance of the Area*

A country may have importance because it is near a superpower or it controls important transport routes. A conflict in this country will rapidly develop into a flashpoint because of its strategic importance.

2 *The Economic Importance of the Area*

A country may have importance because of its valuable resources needed by other countries. Or it may have important industries owned by other countries. Any conflict in this country will develop into a flashpoint because of its economic importance.

3 *The Political Importance of the Area*

A country may have importance because of its political system which may be feared by other countries. A **communist** or **left-wing** take-over may be feared by **capitalist** or **right-wing** countries and vice versa.

Many people believe that when one country is taken over by a left-wing or right-wing government, its neighbours soon follow. This is called the **domino theory.**

4 *The Cultural Importance of the Area*

A country may have importance because it has close cultural ties with other countries eg countries with similar races, history, religions. Any conflict in this country will rapidly develop into a flashpoint because of its cultural importance.

16D The United Nations Organisation

Look back at page 72 to remind yourself of some of the agencies of the United Nations.

The United Nations has 159 member countries. It has three major aims:

1 To achieve social and economic progress
2 To maintain world peace
3 To protect human rights.

We found out in Unit 14 how the United Nations tried to achieve social and economic progress. Maintaining world peace is the task of the United Nations Security Council. When a conflict occurs it has certain powers:

(*a*) Member nations can impose economic sanctions (stop trade) on the offending country.
(*b*) Member nations can impose an arms embargo on the country.
(*c*) It can send in a peacekeeping force, from the member nations.
(*d*) It can send in military observers, to patrol ceasefire agreements.

Unfortunately, in many conflicts, the United Nations cannot help because

1 Any member nation can veto (say no to) any action the UN might want to take.
2 It cannot force member nations to carry out its decisions.

Questions

Look at 16C.

E1 Explain what is meant by a country's (*a*) strategic importance, (*b*) economic importance, (*c*) cultural importance and (*d*) political importance?

E2 What is the 'domino theory'?

Look at 16D.

E3 What are the three major aims of the United Nations Organisation?

E4 Describe the measures the UN can use to stop conflicts.

E5 Describe the difficulties the UN faces in implementing these measures.

Questions

Case Study of Vietnam, South-East Asia

Look at fig. 16.18.

C1 To what extent have the changes in government in South-East Asia since 1948 fitted the domino theory?

Look at fig. 16.19.

C2 Compare the reasons for the involvement of China, USA and USSR in Vietnam since 1961.

Look at figs. 16.21 and 16.24.

C3 Describe the different points of view people in the USA would have had towards the US involvement in Vietnam (1961-1975).

Look at fig. 16.22.

C4 Compare the attitude of the USA towards involvement in other countries before and after the Vietnam War.

Look at fig. 16.22

C5 None of the United Nation's solutions for peace in Vietnam were successful. Suggest reasons why each of the United Nations solutions failed.

UNIT 17 Alliances

Core Text

17A Alliances

Every country in the world needs help from other countries. They may need other countries to protect them or to trade with or to provide them with aid. For these reasons most countries in the world have grouped themselves together. Groups of countries that work together are called **alliances.** Each member country is called an **ally.**

17D Defence Alliances

Some countries group together as alliances for defence. A country is less likely to be attacked if it has allies which will help to defend it. The two biggest defence alliances are NATO and the Warsaw Pact. NATO (North Atlantic Treaty Organisation) was set up in 1949 to protect the capitalist countries of Western Europe from the Soviet Union and the communist countries of Eastern Europe. The Warsaw Pact was set up in 1955 to protect Eastern Europe and the Soviet Union from Western Europe.

Core Questions

Look at 17A.

1 What is an alliance?

Look at 17D.

2 Why do countries form defence alliances?

Look at 17B.

3 Why do countries form trade alliances?

4 Why do outside countries find it difficult to sell goods in a trade alliance?

17B Trade Alliances

Some countries form alliances to make it easier to trade with each other. Each country in a trade alliance is allowed to sell its goods in all the other member countries. But if outside countries want to sell their goods, a tariff (a tax on the price) or a quota (a limit) is put on their goods.

The United Kingdom belongs to a trade alliance called the European Community. COMECON is a trade alliance for the countries of Eastern Europe and the Soviet Union.

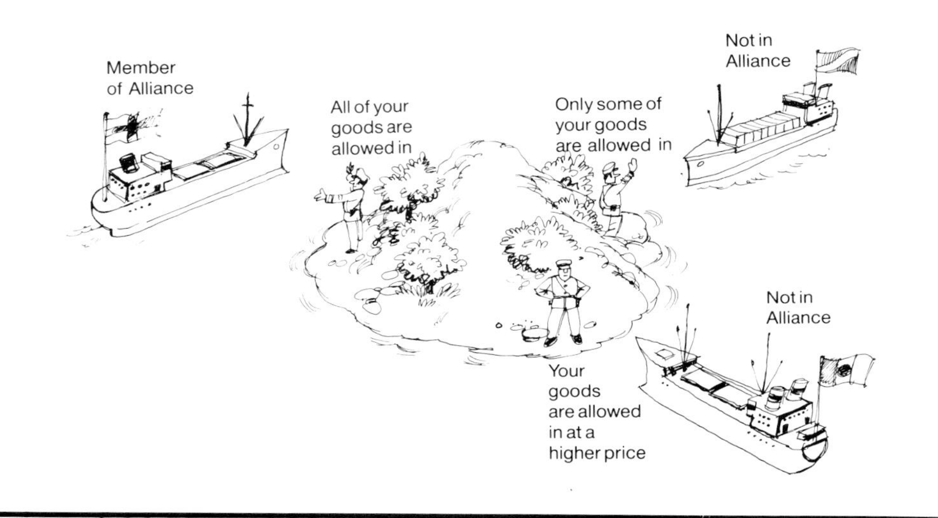

17C Selling Alliance

Countries which sell the same goods can get a higher price if they form an alliance and agree to sell at the same price.

An example of a selling alliance is OPEC (Organisation of Petroleum Exporting Countries). These countries (eg Saudi Arabia, Iran, Kuwait) formed an alliance to sell oil at a high price eg in 1973 they raised its price by 400 per cent. Other countries had to pay up because they could not get their oil from anywhere else.

A SELLING ALLIANCE

Look at 17C.

5 Why do countries form selling alliances?

	NATO	Warsaw Pact COMECON	OPEC	European Community
Algeria			✓	
Belgium	✓			✓
Bulgaria		✓		
Canada	✓			
Czechoslovakia		✓		
Denmark	✓			✓
Equador			✓	
Eire	✓			✓
France	✓			✓
Gabon			✓	
Germany	✓			✓
Greece	✓			✓
Hungary		✓		
Iceland	✓			
Indonesia		✓		
Iran			✓	
Iraq			✓	
Italy	✓			✓
Kuwait			✓	
Libya			✓	
Luxembourg	✓			✓
Netherlands	✓			✓
Nigeria			✓	
Norway	✓			
Poland		✓		
Portugal	✓			✓
Qatar			✓	
Romania		✓		
Saudi Arabia			✓	
Spain	✓			✓
Turkey	✓			
USSR		✓		
United Arab Emirates			✓	
UK	✓			✓
USA	✓			
Venezuela			✓	

Look at the table above.

6 Which countries are members of (*a*) NATO (*b*) Warsaw Pact and Comecon (*c*) European Community (*d*) OPEC?

7 Which two of the 5 alliances in the table are defence alliances?

8 Which two are trade alliances?

9 Which one is a selling alliance?

Questions

Case Study of Jamaica, West Indies

Look at figs. 17.6 and 17.8.

F1 Jamaica belongs to alliances for selling sugar and bauxite. Which is more important to Jamaica – the sugar alliance or the bauxite alliance? Give a reason.

'The International Bauxite Association has been a great help to Jamaica'.

Look at fig. 17.7.

F2 Do you agree with the point of view above? Give a reason for your answer.

Look at figs. 17.5 and 17.9.

F3 Jamaica is a member of the Organisation of American States. Do you think other countries are less likely to attack Jamaica now? Give a reason for your answer.

Look at fig. 17.4.

F4 Do you think the agreement between Jamaica and the European Community is good for Jamaica? Give a reason for your answer.

Look at figs. 17.2 and 17.3.

F5 Which trade alliance should help Jamaica more – CARICOM or CBI? Give a reason for your answer.

F6 Jamaica belongs to a trade alliance, a selling alliance and a defence alliance. Which is the most useful alliance to Jamaica? Give reasons for your answer.

JAMAICA'S ALLIES
Trade Allies
Bahamas
Guyana
Trinidad and Tobago
USA
and Others
Selling Allies
Australia
Dominican Republic
El Salvador
Guinea Guyana
Surinam and Others
Defence Allies
Argentina
Barbados
Brazil
Dominican Republic
Mexico
Surinam
Trinidad and Tobago
USA and Others

Look at the table above.

F7 Which six countries are Jamaica's best allies? Give reasons for your answer.

Questions

Case Study of Benin, West Africa

Look at fig. 17.10.

G1 Compare the changes in the price of palm oil and cotton.

Look at fig. 17.13.

G2 Do you think the International Cocoa Organisation has helped Benin? Give reasons for your answer.

Look at figs. 17.10 and 17.13.

G3 Which selling alliance would be more useful to Benin – an alliance for selling palm oil or for selling cotton?

Look at figs. 17.12 and 17.14.

G4 Give one advantage and one disadvantage of the ACP alliance to Benin.

Look at figs. 17.15 and 17.16.

G5 Benin is a member of the Organisation of African Unity. Do you think other countries are less likely to attack Benin now? Give reasons for your answer.

Look at figs. 17.10 and 17.11.

G6 People in Benin have different opinions over the usefulness of the West African Common Market. Describe two different points of view that people in Benin might have.

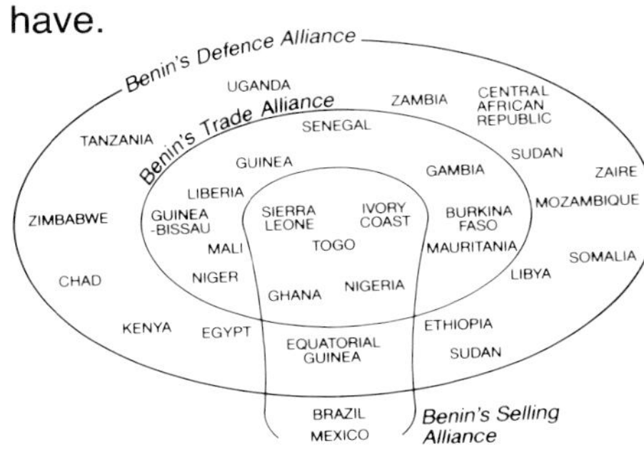

Look at the venn diagram above.

G7 Which countries are Benin's best allies? Give reasons for your answer.

Resources

Case Study of Jamaica, West Indies

Fig. 17.1

Jamaica's Trade Alliances

Jamaica is a member of CARICOM – a group of small, poor Caribbean countries. Jamaica is also in the CBI (Caribbean Basin Initiative). This allows her to export goods to the USA without tariffs.

Fig. 17.2

Where Jamaica's Exports Go (1985)

USA	35%
UK	19%
Canada	18%
Caricom countries	12%
USSR	6%

Fig. 17.3

Jamaica and the European Community

In 1980 the European Community allowed Jamaica to export an agreed amount of its goods to the European Community. The price for these goods was set by the European Community and not by Jamaica.

Fig. 17.4

Jamaica's Defence Alliance

Jamaica has only 3000 armed forces but she belongs to the OAS (Organisation of American States). The countries in this alliance try to achieve peace and justice and defend the territories of each member country.

Fig. 17.5

Jamaica's Selling Alliances

Jamaica belongs to the International Sugar Organisation and the International Bauxite Agreement. These alliances try to get higher prices for sugar and bauxite.

Fig. 17.6

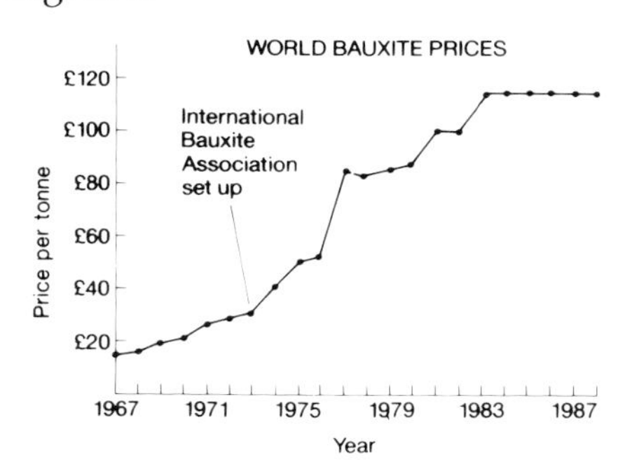

Fig. 17.7

Jamaica's Exports (1985)

1 Alumina	35%
2 Bauxite	15%
3 Sugar	10%

Fig. 17.8

Fig. 17.9

Case Study of Benin, West Africa

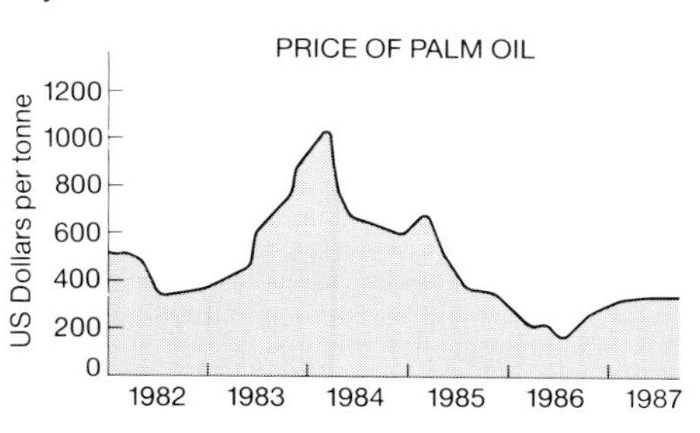

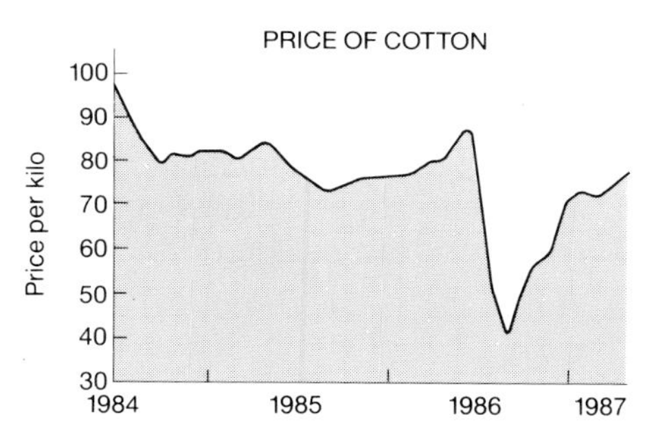

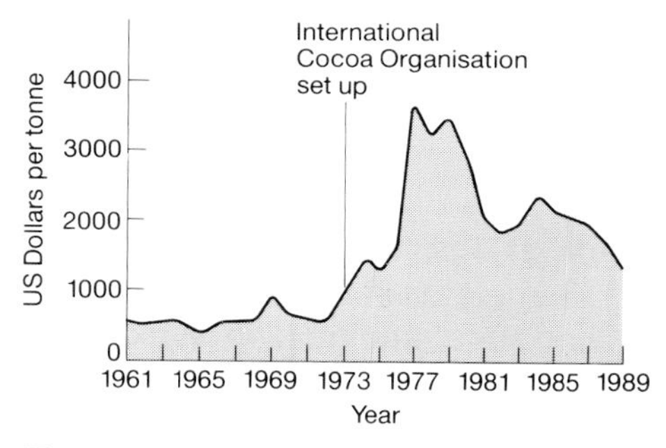

Fig. 17.10

Benin's Trade Alliance
Benin is one of 16 countries in ECOWAS (West African Common Market). The aim of the alliance is that members can export to each other without tariffs. Outside countries have tariffs put on goods they wish to sell in the Common Market.

Fig. 17.11

Where Benin's Exports Go (1985)	
1 Spain	32%
2 West Germany	21%
3 France	16%

Fig. 17.12

Benin's Exports (1985)	
1 Palm oil	32%
2 Cocoa	29%
3 Cotton	24%

Fig. 17.13

Benin is an ACP member of the European Community. This means that she can export a fixed amount of goods (a quota) to the European Community at an agreed price each year.

Fig. 17.14

Benin has only 6000 Armed Forces but she is a member of the OAU (Organisation of African Unity). This organisation aims to unite its member countries and co-operate in matters of defence.

Fig. 17.15

Fig. 17.16

Case Study of Australia

Australia's Exports (1989)	
1 Coal	15%
2 Minerals	14%
3 Wool	11%
4 Cereals	8%

Fig. 17.17

The Price of Wool (in US Cents / kilo)	
1977	360
1979	380
1981	440
1983	500
1985	540

Fig. 17.18

World Producers of Wool (1989)	
1 Australia	25%
2 USSR	16%
3 New Zealand	13%
4 China	6%
5 Argentina	5%

Fig. 17.19

Australia's Defence Alliances
Australia was a member of SEATO (South-East Asia Treaty Organisation) with other countries in South-East Asia. Its aim was to defend the area against aggression from other countries. This has been disbanded and Australia now belongs to ANZUS with the USA.

Fig. 17.20

Some Developments in South-East Asia

1945	North Korea becomes communist
1949	China becomes communist
1955	North Vietnam becomes communist
1969	Communist rebels fighting in the Philippines
1975	All of Vietnam becomes communist. Cambodia becomes communist. Laos becomes communist

Fig. 17.21

Where Australia's Exports Go 1964	
1 UK	20%
2 Japan	17%
3 USA	10%
4 China	5%
5 New Zealand	5%
6 France	4%
7 Malaysia	4%
1987	
1 Japan	25%
2 USA	12%
3 New Zealand	6%
4 China	4%
5 North Korea	4%
6 UK	4%
7 Taiwan	3%

Fig. 17.22

1961 France joins the European Community
1966 Free trade alliance for some goods between New Zealand and Australia
1967 Malaysia joins the South-East Asia Trade Alliance
1973 UK joins the European Community

Fig. 17.23

Sheepfarming in Australia

Australia's Social Alliance
Australia is a member of the Colombo Plan, which gives aid to develop the countries of South-East Asia. It is also one of 50 countries in the British Commonwealth.

Fig. 17.24

Extension Text

17E Social Alliances

Some countries have grouped together for social reasons. They co-operate in many ways eg trade, defence, and aid and so gain many benefits from their alliance.

The British Commonwealth is a social alliance of countries united by an allegiance to the Crown and to the Queen as Head of the Commonwealth. The Arab League unites countries with a common religion, language and cultural heritage.

17F Problems of Defence Alliances

When countries group into defence alliances they become much more powerful. Their strength can be threatening to other defence alliances, which then build up their own military strength to achieve a **balance of power.** This, in turn, can lead to an **arms race** as the alliances increase their military strength to keep up with or a little ahead of other alliances.

NATO and the Warsaw Pact were involved in an arms race for many years. This has led to a build up of both conventional and nuclear weapons on both sides. They now spend 80 per cent of all money spent on arms in the world.

17G Problems of Economic Alliances

There are now many trade alliances throughout the world. They make it easy for each country to trade within its alliance but much more difficult to trade with countries in another alliance. Third World countries, which are often in alliances with each other, find it increasingly difficult to export to the richer countries.

The countries of Western Europe are among the richest in the world. By forming the European Community, it is now more difficult for other countries to trade there. This deprives poorer nations of valuable markets. Because of this an agreement, called the Lomé Convention, between the European Community and the ACP countries (African, Caribbean and Pacific) allows the ACP countries to import agreed amounts at agreed prices into the European Community.

Special trading arrangements can help Third World countries but, in general, the formation of trade alliances has hindered their development. The establishment of selling alliances for all primary goods, leading to higher prices, would do most to help the Third World countries. Unfortunately, with the exception of OPEC, these have proved difficult to achieve.

Questions

Look at 17E.

E1 In what ways do social alliances differ from other alliances?

Look at 17F.

E2 What is meant by the 'balance of power'?

E3 Explain how the formation of defence alliances can lead to an arms race.

Look at 17G.

E4 Explain how the establishment of trade alliances has affected Third World countries.

E5 Explain the purpose of the Lomé Convention.

Questions

Case Study of Australia

Look at figs. 17.17, 17.18 and 17.19.

C1 Describe the arguments for and against Australia setting up an alliance for the selling of wool.

Look at figs. 17.20, 17.21 and a map of Australia and South-East Asia.

C2 Which defence alliance would be better for Australia – an alliance with some countries of South-East Asia or an alliance with the USA? Give reasons for and predict the consequences of your decision.

Look at figs. 17.22 and 17.23.

C3 Describe the relationship between Australia's changing export pattern and the establishment of international trade alliances.

Possible Alliances Between Australia and South-East Asia	
1	Social alliance (the Colombo Plan) – involves Australia giving aid to South-East Asia.
2	Trade alliance – allows free trade between South-East Asia and Australia.
3	Selling alliance – for all primary goods that the countries produce.

Look at figs. 17.22 and 17.24.

C4 Which type of alliance between South-East Asia and Australia, in the table above, would be most helpful to the countries of South-East Asia? Give detailed reasons for your answer.

UNIT 18 The European Community

Core Text

18A The European Community

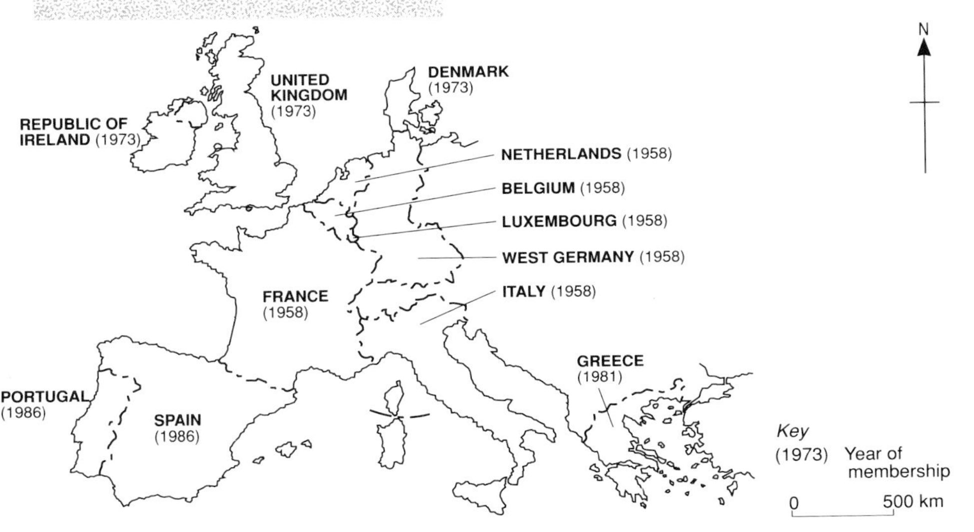

The European Community is an alliance between the 12 countries shown above. It began with 6 countries in 1958. Since then 6 more countries including the UK have joined.

Like all alliances, the countries of the European Community have grouped together to get benefits. There are in particular benefits to trade, farming and industry.

	European Community	USSR	USA
Area (millions sq. km)	2	22	9
(1986) Population (millions)	323	280	242
(1986) Wealth (£ billion)	1250	400	1900

18B The European Community and Trade

The European Community helps its countries to trade with each other because it has removed all tariffs (taxes) and quotas (limits) on goods going between the countries. It also helps them by making it more difficult for other countries to sell their goods in the European Community. It does this by putting on tariffs, which raise the price of their goods.

18C The European Community and Farming

The European Community helps its farmers by giving them **guaranteed prices** for their products. These are prices which are high enough for all farmers to earn a good living. If the farmer cannot sell his products to anyone else, the European Community itself will buy them from him at the guaranteed price. This is the **Common Agricultural Policy** (**CAP**). It makes farming less risky.

Sometimes the guaranteed prices are too high. Then the farmers produce so much that not all of it can be sold. This means there is a **surplus** (more than is needed). Then the European Community either stores it or sells it to other countries.

18D The European Community and Industry

The European Community helps industries in those regions which suffer from high unemployment and poverty. The European Community gives **grants** to companies which set up in these regions. It also gives grants for other things which will help the regions eg new roads. It may pay for the local people to be trained in new skills. This helps the industries and the people that live in these regions.

By improving trade, farming and industry, the European Community improves the standard of living of the people in its member countries. In return, each country has to pay money into the European Community Fund each year.

Core Questions

Look at 18A.

1 Name the 12 countries of the European Community.

Look at 18B.

2 In what ways does the European Community help its countries to trade with each other?

3 Why do other countries find it more difficult to sell goods in the European Community?

Look at 18C.

4 In what ways does the European Community help farmers?

5 Why does the European Community sometimes have surpluses of butter, grain and beef?

Look at 18D.

6 In which regions does the European Community help industries?

7 What help does the European Community give to industries in these regions?

Questions

Case Study of Scotland

F1 What does fig. 18.2. show?

Look at the headlines below.

F2 Do you think that Scottish electronics companies were pleased when we joined the European Community?

F3 What does fig. 18.4. show?

Look at fig. 18.3.

It shows four schemes in Scotland funded by the European Community.

F4 Which out of the four schemes has been of most help to the local people? Give reasons for your answer.

'Unemployment here in the Garnock Valley is 10 per cent. The European Community hasn't helped us much.'

Look at fig. 18.5.

F5 The statement above is exaggerated. Explain how it is exaggerated.

Questions

Case Study of South Italy in the European Community

G1 What does fig. 18.9 show about Italy and the European Community Fund?

Look at fig. 18.8.

G2 Joining the European Communuity has brought advantages and disadvantages to shoe manufacturers in Italy. Give one advantage and one disadvantage.

G3 Which two of the facts in fig. 18.7. best indicate that industries in south Italy needed help from the European Community in 1960? Give reasons for your answers.

Look at fig. 18.12.

G4 Compare the effects of joining the European Community on farmers growing tomatoes and olives in south Italy.

Look at fig. 18.10.

G5 Many people in south Italy are happy with the developments in farming. Some people are not happy.
Describe two different points of view towards the developments in farming in south Italy.

Look at fig 18.11. and the cartoon below.

G6 What is the cartoon trying to show?

G7 Which of the schemes shown in fig. 18.13. will have helped the people of south Italy the most? Give reasons for your answer.

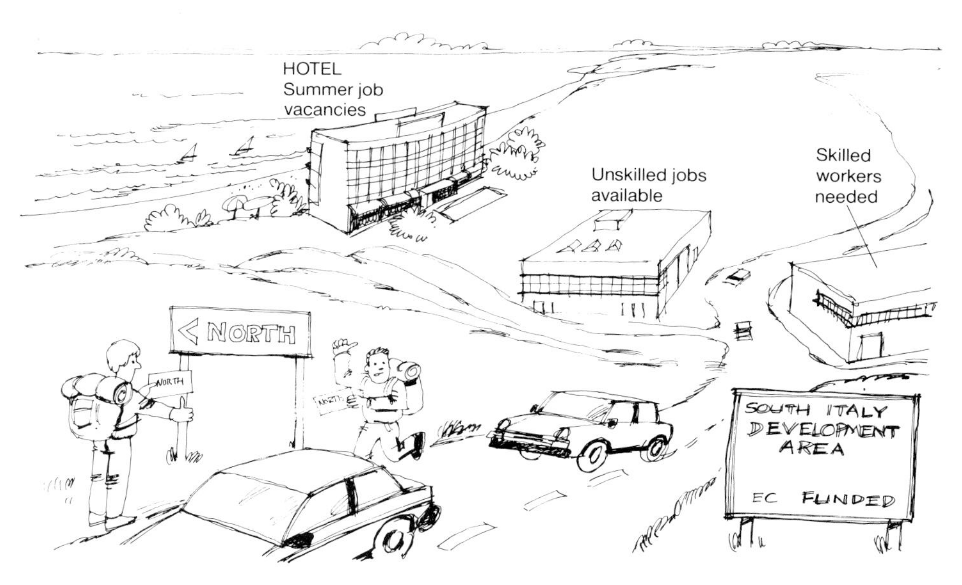

Resources

Case Study of Scotland in the European Community

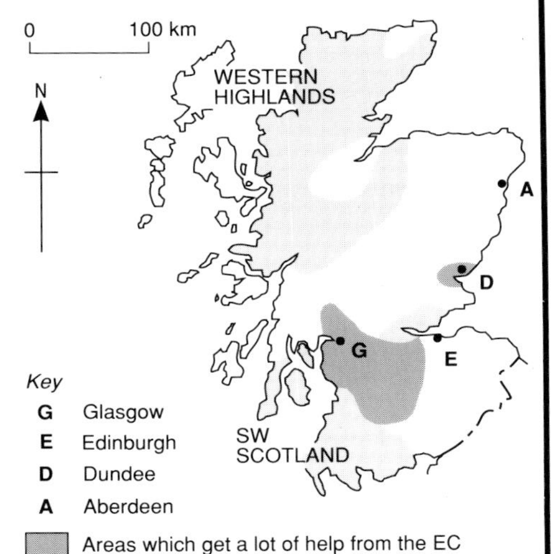

Fig. 18.1

In 1987, the UK contributed £5700 million pounds to the EC fund and in the same year received £4500 million in return.

Fig. 18.2

Some EC Funded Schemes in Scotland

1 New leisure centre in Arbroath.
2 New factory in Glasgow employing 200 people.
3 New power station in the Orkney Isles, providing cheap electricity.
4 New Edinburgh City by-pass.

Fig. 18.3

	Wheat Prices per tonne (1985)
Price EC farmers get	£216
Price other farmers get	£162

Fig. 18.4

Garnock Valley, Strathclyde

When the steelworks in the Garnock Valley closed, unemployment rose to 33 per cent. The European community gave grants and loans for building factories. Chemical works, glassworks and car plants moved in. The Community also paid small new companies £20 per week for every worker they employed. Unemployment in the Garnock Valley had dropped to 10 per cent by 1990.

Fig. 18.5

Case Study of South Italy in the European Community

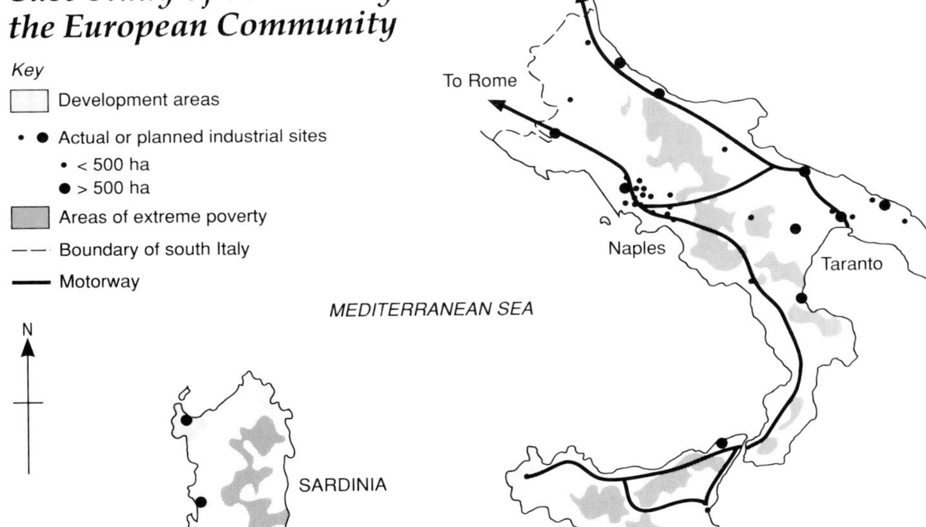

Fig. 18.6

Facts about South Italy (1960)

Population 20 million;
150000 emigrants each year;
40 per cent work in agriculture;
enormous farm estates;
5 months of summer drought;
80 per cent is steep mountains;
Income per person half the European Community average;
Frequent earthquakes and volcanic eruptions;

Fig. 18.7

The Italian Shoe Industry

Before Italy joined the European Community, Italian shoes sold in Britain had an 8 per cent tariff added to their price. This made it more difficult to sell them. Now there is no tariff.

Before Italy joined the European Community, all British shoes had a 10 per cent tariff added to their prices before they could be sold in Italy. Now there is no tariff.

Fig. 18.8

Italy's population comprises 17 per cent of the total population of the European Community and yet it receives 19 per cent of the funds allocated by the EC. In return, it contributes 13 per cent of the total contributed by member states.

Fig. 18.9

The European Community and Farming in South Italy (1960-1984)

The huge farm estates were broken up and new farms were created of between 5 and 50 hectares in size. Some proved to be too small to be profitable. 100 000 extra families received farms, but many more received no farms at all. Grants were given for 20 000 wells to be dug for irrigation. Now farmers can grow crops such as fruit and salad crops, which fetch high prices. Some farmers, however, have found it difficult to change to new farming methods.

Fig. 18.10

The European Community and Industry in South Italy

The European Community and the Italian government provided grants for building new factories, hotels and roads. Steelworks, engineering works, oil refineries and many others have all moved to the south and over 500 000 jobs have been created. But many jobs proved unsuitable for the local people who had only worked on farms before. Many jobs in tourism were for the summer season only. Many of the large factories use few workers and most of the well paid jobs have gone to people from the north. Now grants are being given more to local medium sized companies.

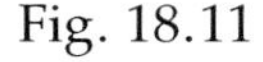

Fig. 18.11

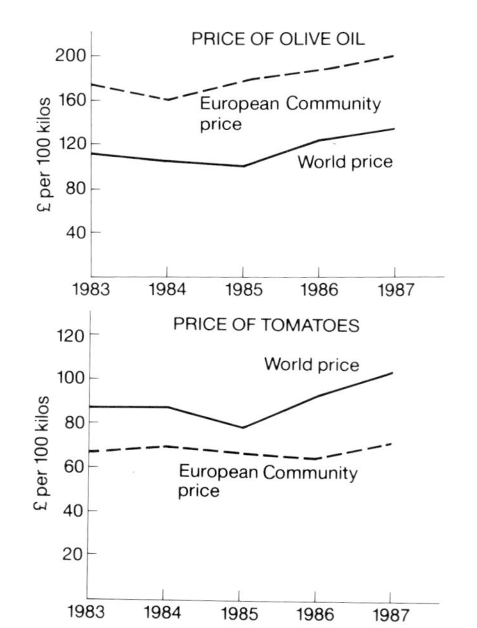

Fig. 18.12

Some European Community Funded Schemes in South Italy

1 A huge estate in Basilicata broken up and 5000 new farms created, all with irrigation water.
2 A new steelworks in Taranto, employing 30 000 people.
3 A new motorway to connect south Italy with the north, so that farm produce and factory goods can get to their markets quickly.
4 A cattle breeding station in Catania, where new breeds of dairy cattle, which can withstand the local climate, are bred.

Fig. 18.13

Case Study of the Irish Republic

	Population millions	Unemploy-ment (%)	GNP/head (dollars)	Net receipts from EC (%)
Belgium	10	11	9230	− 4
Denmark	5	6	12600	− 1
France	55	10	10720	− 9
Greece	10	8	3680	+ 8
Ireland	4	18	5070	+ 4
Italy	57	11	8550	+ 9
Luxembourg	0.4	2	14260	0
Netherlands	15	10	10020	− 6
Portugal	10	5	2250	+ 9
Spain	39	19	4860	+ 8
UK	57	9	8870	+ 4
W. Germany	61	6	12080	− 22

Fig. 18.14

	Population (1987)	Unemployment (%)	In agriculture (%)
Connaght	431 000	19.8	25.5
Leinster	1 851 000	17.1	9.1
Munster	1 020 000	17.2	21.3
Ulster (part of)	2 36 000	19.2	20.0

Fig. 18.15

Views on using EC Funds

'We should give funds to each province according to their population'
'Better to give funds according to their unemployed rate'
'We could give funds according to the percentage of people in farming'
'Why don't we give all the funds to the poorest province?

Fig. 18.16

Ireland's Export Partners

1970
1 UK
2 USA
3 France
4 West Germany
5 Netherlands
6 Belgium/Luxembourg
7 Canada

1973
1 UK
2 USA
3 West Germany
4 France
5 Netherlands
6 Belgium/Luxembourg
7 Italy

1975
1 UK
2 West Germany
3 Netherlands
4 France
5 Belgium/Luxembourg
6 Italy
7 USA

Fig. 18.17

Some European Community Funded Schemes in the Irish Republic

1 A regional airport in Munster
2 A technical college in Leinster
3 An engineering works in Leinster, employing 500 people
4 A textile works in Connaght, employing 500 people

Fig. 18.18

Extension Text

18E European Community Funding Problems

The European Community has brought many advantages to its member countries. It has also created problems. The richer countries, such as Germany give more money to the European Community Fund than they receive from it. They are called **donor nations**. Others, such as Portugal and Greece, receive more than they give and are called **receiver nations**. This has caused many arguments among the members.

18F European Community Trade Problems

Increased trade between the countries has meant that they trade less with other countries. Before the European Community was set up many of the countries traded a lot with their old colonies – Third World countries in Asia, Africa and the Caribbean. These countries have suffered badly. The European Community has tried to resolve this problem by the Lomé Convention. Sixty ACP countries (African, Caribbean and Pacific) are allowed to sell most of their exports in the European Community without tariffs. The Community also provided these countries with £550 million of aid between 1986 and 1990.

18G European Community Industry Problems

Abolition of tariffs have made it easier for companies to sell their goods in other member countries, but it has led to increased competition at home. This has led to lower prices, but some companies have had to close as a result.

Grants from the European Community have helped regions with high unemployment, but the regions were so large that the funds were spread over too wide an area. These regions have now been reduced in size to concentrate help in the areas most in need.

18H European Community Farming Problems

High guaranteed prices for cereals in the European Community have led to many changes in the landscape.

High guaranteed prices have also led to surpluses. Farmers have even been paid not to harvest their crops. To store surplus wheat cost over £3000 per minute in 1987. This is at a time when 2000 million people in the world do not get enough food to eat. Indeed for every £1 that the Community spends on aid, it spends £3 in maintaining its surplus food.

EUROPE'S CHANGING COUNTRYSIDE

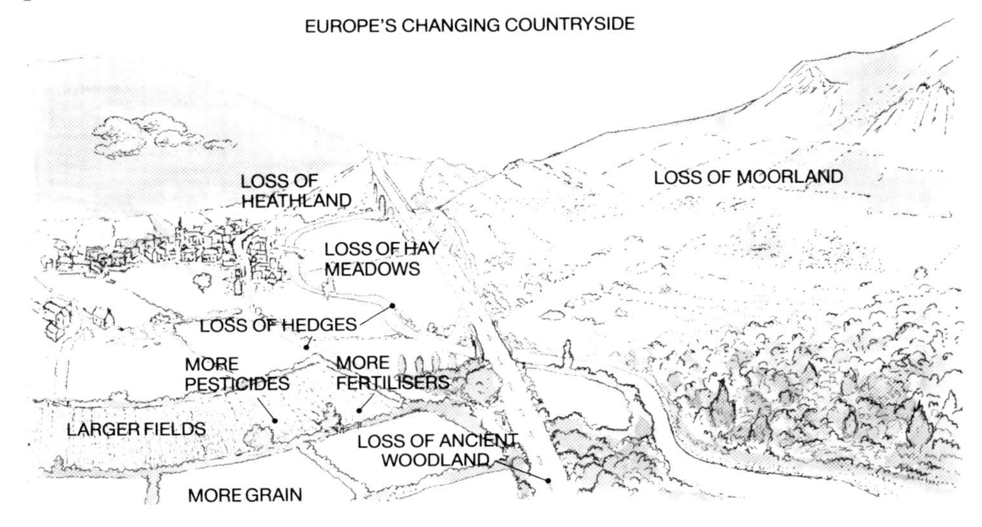

Questions

Look at 18E to 18H

E1 Explain the meaning of a 'donor nation'.

E2 Describe the effects of high grain prices on the European Community.

E3 Explain how the abolition of tariffs has led to lower prices.

'Giving grants to regions of high unemployment is like irrigating the desert with a watering can'

E4 Explain what the statement above means.

Questions

Case Study of the Irish Republic

'Ireland has a small population so it should only get a small amount of money from the European Community Fund' (German minister)

Look at fig. 18.14.

C1 The statement above is biased. Explain how it is biased.

Look at figs. 18.15 and 18.16.

C2 Describe the arguments for and against the different points of view in fig 18.16.

Look at fig. 18.17.

C3 Describe the changes in Ireland's export partners before and after Ireland joined the E.C.

Look at figs. 18.15 and 18.18.

C4 Which European Community funded scheme will have helped the Irish Republic most? Give detailed reasons for your answer.

Index